Figures for Fun

Aldine Paperbacks

Figures for Fun

J.A.H.Hunter

J.M.Dent & Sons Ltd

Printed in Great Britain by
Lowe & Brydone (Printers) Ltd
for
J. M. Dent & Sons Ltd
Aldine House · Bedford Street · London

First published 1957
Second impression 1963
Revised edition 1972

ISBN 0 460 02095 1

CONTENTS

Preface

Even the real mathematician may find something to amuse him in this little selection of teasers, but it is intended mainly for the more normal people who, like myself, take figures rather less seriously.

For figures *can* be fun, as seems to be confirmed by the vast interest taken in such problems by thousands of kind readers, without whose ideas, encouragement and support this little book might never have been published.

So now I take this opportunity of saying 'Many thanks' to all those good friends, known and unknown.

London, 1956 J. A. H. HUNTER

Preface to Aldine Paperback Edition

The success of my original *Figures For Fun* in hardback form has prompted publication of this edition for an even wider distribution.

The decimal currency, and the introduction of a decimal system of weights and measures that is planned for 1975, have necessitated some changes in the content.

Since the original book was published the alphametic has become immensely popular throughout the world as a type of strictly mathematical puzzle that requires only a good knowledge of basic arithmetic coupled with sound reasoning. Two more puzzles of this type have been included in the content.

Otherwise, there have been no changes.

London, 1971 J. A. H. HUNTER

FIGURES FOR FUN

1. Murder on the River

'All aboard!' called Mary, stepping gingerly into the canoe from the slippery steps at the foot of the garden. She settled herself into the cushions while her husband started to paddle upstream; she yawned, she slept.

The noise of traffic overhead wakened her as they passed under the bridge some minutes later. She opened her eyes, looked over the side, and screamed: 'Steve, there's a hand in the water! It's a dead body!'

'You've been dreaming, darling,' he laughed, seeing what she saw. 'It's only an old glove full of nothing.' Relaxing with a sigh of relief, Mary closed her eyes again, and Steve went on paddling doggedly upstream.

After a further fifteen minutes he turned and made for home. 'Whoosh, whoosh, whoosh,' went his paddle—always the same steady rhythm as the light shell slipped through the water. They shot under the bridge, and then it didn't take long to make the remaining mile to their house.

As they arrived abreast their steps, Steve roused his wife with a shout: 'Wakey, wakey! Here we are, and there's your corpse.' And there indeed, still floating in mid-stream, was that very lifelike old glove which they had just overtaken.

Steve paddled at the same speed all the time. But what do you make the speed of the current?

2. The Car Game

Jack and Jill sat at the big bay-window, playing the car game. She paid him a penny for every car that passed from right to left, and he paid her the same for every car passing the other way.

They live in a quiet road with little traffic, and most of the cars seemed to be passing in Jack's favour. 'It isn't fair,' complained Jill at last, 'only seven have gone my way.'

'All right;' replied Jack, who was becoming rather bored with the game, 'we'll stop after five more have passed'. But, of those next five cars, only one went his way and so he ended up only one penny to the good.

So how many cars passed during the game?

3. Peace at a Price

'All right! I'll play with you, but only for a few minutes,' said Uncle Jim, knowing very well his nieces would give him no peace unless he did so. He jingled a fistful of coins and promised them to the one who guessed their total value.

'Twenty-three pence,' hazarded Pam. After some more jingling, Pat took a chance. 'Ninety-eight pence,' she suggested. Peg was quite confident. 'Just eight pence,' she cried, 'so pay up and no cheating!'

But Uncle Jim opened his hand and of course they were all wrong. He saw his chance to buy peace, however. 'Pam was nearest,' he said, 'so I'll give her the prize if you'll all leave me alone for the rest of the afternoon.' And Pam saw to it!

In fact, one of their guesses was twice as much in error as one of the other two, and that in turn was twice as much in error as the third. So how much was there in his hand?

4. The Spider and the Fly

'Come right into my parlour,' said
The spider to the fly,
'And answer one small question, please,
Unless you want to die.
I've eaten scores of flies, of course,
But tell me if you dare:
If females had two more, and males
But half their present share,
How many flies like that, d'you think,
I really would require,
To give me twenty-eight fly legs,
The number I desire?'

5. Such a Waste

'What happened, dear?' asked John, as his wife came into the sitting-room, laden with parcels and obviously upset.

'Everything! The market was a bedlam, and then I had to fight my way into the bus and my bag burst and I lost all my oranges.' Gwen sounded very near to tears, and John tried to console her: 'What's a few oranges, anyhow?'

'But it's such a waste, the price things are these days,' she sighed, 'and they were lovely ones at five pence each.'

It would have been more serious if she had lost the peaches, of which she had bought more and paid as many pence each as the number she bought. As Gwen paid 89p altogether for the oranges and peaches, you can probably work out how many oranges she lost.

6. If and If,

'Here's ten pence for each year of your age, and five pence for each year of Jill's,' said Uncle Frank, handing Jack some coins. 'Get yourselves some Easter eggs.'

The boy started to thank him, and then thought of something. 'If I were a year older than I am and Jill a year younger than she is, but you gave us five penny pieces for my age and ten-penny pieces for Jill's, I wonder what would happen.'

'I know.' His uncle chuckled. 'You'd be getting just fifteen pence less.'

How much older is Jack than his sister?

7. The Funny Figures One Sees

Bert and Ben were strolling down Edgware Rd., when Ben stopped suddenly. 'Look at those funny figures,' he exclaimed, pointing across the street.

'Don't see anything funny about them,' rejoined his friend, 'they look all right to me!' 'Not those, you old wolf;' laughed Ben, 'I mean the number of that shop.'

Bert dragged his eyes away reluctantly, and waited: he guessed what was coming. 'If you divide it by three,' continued Ben, 'there's one over; by five there's two over; by seven there's three, and by nine there's four over.'

'What's funny in that? There must be lots of numbers like that with this street running miles up north:' Bert was in no mood for such nonsense.

Ben nodded: 'Yes, of course; but we're only in the one-hundreds here.'

Bert wasn't being co-operative, but perhaps you can find that number.

8. Trust a Postman to Know

Myrtle Drive is a pleasant road of modern villas, seldom used by through traffic: perhaps its only fault is that the houses are all alike, but that's merely a matter of opinion. Each doorway faces its twin squarely across the intervening hundred-and-twelve feet of turf and road. By common agreement, the owners of these houses have put up no fencing between them to the front, and so a broad swathe of lawn runs right along each side between houses and pavement, broken only by neat flagged paths.

The postman comes to this road from the east: after Myrtle Drive he delivers in Sylvan Avenue, which runs across its western end. Today, Mrs Smart, who lives at No. 7 and always watches for her mail, noticed the postman was following a different route with his letters. When he reached her door she mentioned this. 'Yes, Mum,' he replied, 'I reckoned I was wasting shoe-leather in this road, so now I get around the job in the shortest possible distance.'

'But you went right down the other side today, and then back to the beginning of this side by a long diagonal, and now you'll have to finish this side;' observed the lady, 'you ought to do it my way.' She explained her idea: to start at No. 1, cross the road to No. 2 and then call at the neighbouring No. 4, then across to No. 3 followed by No. 5, and so on.

'But the distance would be exactly the same,' the postman told her. And, bearing in mind that each house is built on a frontage of exactly sixty-four feet, Mrs Smart had to agree. So how many houses are there on Myrtle Drive?

9. How Many Cigarettes?

John gave up smoking last month; anyhow, he did cut down considerably on his smoking. He explained that a complete break would be bad for his nerves, but his wife had also heard that one before.

For the first five days, John kept well within the ration he had fixed for himself. The sixth day was bad, however, for friends came in and he had to be sociable: but that day only brought his total for the six days up to his quota.

Next day was Sunday, one of those wet Sundays when you feel there's no place like home. And, going to bed that evening, John was challenged by his wife. 'How's the quota now?' she asked. He had to admit he had smoked as many that day as on the previous day, seventeen cigarettes in fact, and that his daily average was now one over his ration.

So what do you make John's daily quota?

10. No Seven Here

This little teaser is really quite simple, but on second thoughts it isn't too easy at that. Five figures of the multiplication sum are shown, and the rest are indicated by crosses. One hint will be useful: no seven appears in the calculation. So now you may be able to find out what the final result is.

```
    2 x x
    3 x x
    -----
    5 x x
  x 4 x
x x 3
---------
x x x x x
```

11. Tim and his Brother Tom

Tom and Tim are brothers; their combined ages make up seventeen years. When Tom was as old as Tim was when Tim was twice as old as Tom was when Tom was fifteen years younger than Tim will be when Tim is twice his present age, Tom was two years younger than Tim was when Tim was three years older than Tom was when Tom was a third as old as Tim was when Tim was a year older than Tom was seven years ago.

So how old is Tim?

12. Chicken Feed

It was dusk. The last of the chores was done and Bill sat with his wife on the porch, looking out over their farm as the familiar scene melted into the deepening shadows. It was the time they always looked forward to: those peaceful moments together, resting after the day's hard toil.

But Bill was worried: the price of feed was up again. 'A sack of grain lasts our chicks only nine days,' he told Mary, 'and we can't go on like that.' He puffed at his pipe: 'What about getting rid of them altogether?'

Mary's mind flashed to thoughts of his favourite Sunday dinner. 'Surely you'll keep some for ourselves!'

Bill considered the point a moment. 'All right,' he conceded. 'I'll keep a dozen and sell the rest; and we'll cut the ration of grain by ten per cent and that'll make a sack last thirty days.'

How many chickens did they have?

13. No Catch in This

Look! My digits are three,
So just what can they be?
Take a third of my first
  From a half of my third:
And a minus, just here,
  Would of course be absurd.

To continue the riddle,
You must add twice my middle:
If you worked it out right
  Then you can't have got seven;
For, believe me, I know
  That you must have eleven.

Now there's no catch in this,
But don't take it amiss
When I add one thing more,
  Just to make it quite clear:
You must know that a five
  Is no part of me here.

14. Kim's Camera

Kim was showing Joe his new camera, explaining the advantages of coupled rangefinder, coated lens, flash synchronisation and other special features of his purchase. But he was particularly proud of having persuaded the dealer to let him have it at less than the normal price of £50.

'And it was really funny,' he told his friend, 'that I paid an exact number of pounds.'

'Nothing very funny about that,' observed Joe, drily.

Kim smiled. 'But I paid cash,' he explained. 'Pounds, fifty-pence, ten-pence, five-pence, two-pence and one-penny pieces: the same number of each.'

It may have seemed funny to Kim, but Joe still wants to know what he paid for the camera.

15. The Old Soldier

Harry's grandfather is an old soldier: a veteran of many wars, for he took the Queen's Shilling away back in 1890. He was relating some of his more probable adventures, and Harry sat there taking in every word. But, after one particularly lurid episode, the boy interrupted: 'Gee whiz! That was exciting. But how long ago was it?'

The old man took a shilling from his pocket. 'You'll get this if you work out the year;' he replied, 'if you exchange the second and fourth figures you make a date ninety-nine years later than what I'm talking about.'

Harry couldn't do it in his head, but what would you say?

16. Ken's Pocket-Money

First thing every Monday morning Ken receives his ten pence: that's one thing the boy never forgets. By the beginning of last month he had managed to save a bit, but during the month his hoard dwindled away: each week he spent half of what he had after receiving his pocket-money on the Monday.

When it came to the fourth Monday in the month, Ken found that he had only 43p altogether, when he counted his cash after receiving a shiny new ten-pence piece from his mother. It was bad luck that no extras had come his way during the month: he could generally be sure of occasional tips from one or other of his numerous uncles.

How much did he have that first Monday of last month, before receiving his pocket-money?

17. Bicycles and Horses

They had been waiting a long time there in the car, but there was still no sign of Ronnie's mother. Uncle Frank turned to his small nephew: 'We'll play a game,' he said. 'I'll give you five pence for every car that passes, and two and a half pence for every bicycle; and you give me four pence for every horse.'

Traffic was light in the village that day, but it promised to be a profitable deal for the boy. When his mother emerged from the shop, so bringing the game to an end, they had counted only half as many horses as cars. Ronnie didn't want to stop: he had won 48p already.

How many bicycles passed by?

18. Where does the Money go?

Mrs Brown goes shopping most days, and most days she returns home in a complete muddle as to what she's spent. Yesterday, in the first shop she spent half of what she had in her bag. In the next shop she spent half of what remained, and so she then called in at the Bank and cashed a cheque for a couple of pounds.

Out in the street again, however, she found it had started to rain. So she slipped into the chemists across the road and made one final purchase before going home: for that she paid the same amount as she had spent in the first shop.

'Now for a taxi,' she thought, as she emerged from the chemists: there was a veritable downpour outside. But that meant checking her cash to be sure she had enough: all was well, for Mrs Brown found exactly 42p in her bag. 'I'm sure I don't know where it all goes,' she grumbled, hailing a passing taxi.

Of course it's easy to be wise after the event: but how much would you say she had spent?

19. Who's the Lady?

'Saw you with a couple of good-lookers yesterday,' remarked Mike with a sly grin.

'What, Gwen and Peg? Why, one of them's my niece.' Steve hesitated a moment, chuckling. 'When my niece is as old as I am now, Gwen will be twice as old as Peg is now. And I'll gladly introduce you to my niece if you'll tell me her name, even though they're both in their teens.'

'You old wolf!' laughed Mike, who knew very well that his friend was just forty. But he got the introduction.

What do you make of it?

20. How Many Marbles?

'Break it up!' called their father, coming out into the yard where the three boys were playing marbles. 'Dinner's nearly ready!' But he relented when he saw the looks on their faces. 'Okay. Two minutes more, but that's all.'

They continued their game with fierce concentration under his critical gaze. All too soon the time was up. 'Come along now, no more.' And now he meant it.

'When you came out I'd lost as many as Tim had won,' grumbled one of the boys, 'and now we stop just as my luck is turning.'

'But Tim started with half of what I had, and Tam started with five times as many as me,' piped Tom shrilly, 'and now Tam has as many as I had when you first came out.'

The third boy had pocketed his marbles. 'I've got three less than Tam had when we started,' he told them.

'Anyway, two of you ended up equal,' commented their father, leading them into the house.

How many marbles did the boys have among them?

21. A Family Gathering

It was a family re-union dinner: just the six of them. Betty was there with her sister Mrs Armstrong, and her brother, and then there were their respective spouses.

Mr Armstrong sat beside Amy, while Alfred faced Mrs Cross. Charles was placed on the left of Betty's sister, while Betty sat between Herbert and Mr Briggs.

Clare started them laughing when she brought out her little joke: 'There'll be no fighting, as wives and husbands aren't sitting together.' Altogether it was a merry occasion.

Now it may amuse you to work out the first names of those three couples.

22. The Tale of Knok

'There was a man in the land of Kalota whose name was Knok; and he was mighty of stature and a cunning worker withal in brass and copper. And Knok had in his place many stones for the true weighing of his wares: stones of one kalen and other stones of greater weights.

'And it came to pass that in the season of the rains great floods spread across the land, and the people fled to the high places leaving all behind them.

'And after a space, when the waters had receded, the people returned; and there was much lamentation, for the homes of the people were destroyed.

'And in all the rubbish Knok could find only one of his stones, a great stone which weighed forty kalens. And Knok took a mighty hammer and he smote the stone and clave it into four pieces, and no piece was like unto another piece. And with the four pieces was he able to weigh his wares, and the people paid him for each kalen of weight, yea for each kalen from one up to forty kalens.

'So Knok wrought for seven days and seven nights, making for each the wares which he desired, and he waxed rich in the land.'

That is what the Book says. But what did those four pieces of stone weigh?

23. North and South They Go

'Hello, Ken: training for a policemen, eh?' laughed Uncle Ted, coming up to where the boy stood making notes on a piece of paper as the cars streamed by.

Ken checked his figures carefully before answering. 'Seven hundred and eighty cars have passed while I've been here;' he informed his uncle, 'and if a quarter of the cars going north had been going south, and if a fifth of the cars going south had been going north, and if a sixth of the cars which would then have been going north had not been out today at all, there would have been ten more cars going south than north.'

That hardly sounded like a member of the Force, but anyhow Uncle Ted did manage to work out how many cars Ken had counted going north. What do you make of it?

24. Those Licence Plates

Paul was thinking of England, where the licence number of your car doesn't change every year and where you don't have to use drab plates issued by a Government office. He looked down at the tinny new licence plates on the table: the three friends had gone together to renew their licences, and were now having a snack together. 'It's a ridiculous system;' he remarked, 'just as you begin to remember your number they change it.'

'That goes for me too,' agreed Dick, 'but there's something funny I've noticed about our numbers this time. Your first figure is the same as Hal's last, and your last the same as his first, and you've both got the same figures in the middle.'

'And there's something else;' exclaimed Hal, 'your two numbers add up to mine.'

Paul picked up the bill, which his friends seemed too engrossed to have noticed. 'Let's go;' he said, 'but I would have you note that Dick's two middle figures are the same and are also the same as your first figure, and his first figure is the same as mine.'

Well, it was certainly a mix-up of figures between them: fortunately, perhaps, all three had been given four-figure numbers. So now what was Dick's number?

25. So Little for So Many

'Lordy, but I'm tired,' sighed Susie, as she and Sally settled down in the bus. 'The same old skirts every day: about fourteen dozen this week, and I hate the sight of the blessed things.'

'What about me?' said Sally. 'Yours are quick to make and you get eleven pence for each.'

The girls were on their way home that Friday evening from the factory where they are employed on piecework. As the bus jerked its way from stop to stop, they compared notes on the week's work: it transpired that there was a difference of only 10p between their respective earnings for the five days, although Sally was paid 20p for each of the dresses she made.

How much did Susie earn that week?

26. Cigars for Dad

Simple Simon was in a fix: he wanted to get his father a dozen cigars for Christmas, and now he didn't really know which to buy.

'They're *so* long,' he told the assistant, spreading out his fingers, 'with gold bands on them: that's what Dad likes.' The man smiled, but he did help the boy to pick four of a cheap brand, four at double that price, and four at 6p each more than the cheap cigars. But then he found that these would cost him 32p more than he had in his pocket.

In the end he bought a dozen of the medium-priced cigars, and that just took all his money. So how much did Simon spend?

27. A Tale of the Tube

He looked so forlorn, standing there all alone with the heedless, seething crowds milling around him. 'What's the trouble, sonny?' I asked, my heart touched by his sad expression.

'I'm waiting for my brother Bob, but he's awfully late,' replied the boy. Cold shivers rippled down my spine as I envisaged an even smaller lad lost somewhere in the vastness of that tube station. 'That's too bad,' I commented, 'but how old is he?'

'He's big,' was the proud answer, 'as big as you, Sir!' When he spoke again it was in an even more confident tone: 'He's bigger than Mum, but not so old.' I smiled at that, and so must have encouraged the little chap to continue. 'Dad said today that Mum's age in two years' time, divided by Bob's age now and then added to Bob's age, comes to two-thirds of what Mum was when Bob was born.'

He paused a moment, a wicked grin creasing his grubby little face. 'And today's Mum's birthday and also Bob's.'

So this was the little wretch on whom I had wasted my pity, this little demon with his tantalising teaser which still keeps me awake at night! But perhaps you can make something of it.

28. The Birds They Missed

Towards the end of a day's shooting, Jim and Joe were resting on an old stump, enjoying a smoke. Comparing results, they found each had expended the same amount of ammunition: Joe, by no means a good shot, had averaged three cartridges for every bird he had brought down: Jim, who admits he only goes shooting to get away from a week-end at home, had killed only one bird for every four shots.

Suddenly, as they talked together, a flight of duck appeared, coming in low in their direction. Each fired once, and after that they saw no more birds that day.

Trudging back to the inn some time later, tired and very thirsty, Jim remarked on the fact that he had bagged the same number of birds as his friend. 'So what?' chuckled Joe. 'It's also funny neither of us killed two birds with one shot today.'

How many did Jim kill?

29. The Beauties of Kalota

John seems to have had a thoroughly good time last year when he was in Kalota. He did find it difficult, however, to understand the Kalotan women. It is the strange custom of that island that a woman must never make two consecutive true or untrue statements; if one statement is true, then her next is a lie, and vice versa.

The hospitable merchant with whom John was staying had four attractive daughters: Kassa, Kessa, Kissa, and Kossa. John says he tried to find out their ages, but the girls got him really tangled, even though there are no twins in Kalota.

Kassa started it: 'Kissa is twenty-three, and Kessa twenty-two.' Kessa's version was quite different: 'Kossa is twenty,' she told him, 'and Kissa is twenty-two.'

John appealed to Kissa, for whom he appears to have fallen in a big way, but she wasn't very helpful. 'Kassa is twenty-two, and Kossa is nineteen,' she assured him demurely.

John knew very well that one of the girls was nineteen, and so he should have been able to work it all out from what they had said; but then he never was very bright. What do you make their ages?

30. Milly Miffen's Muffins

Milly Miffen made a muffin more than Molly's mother made and Milly Miffen's mother made a muffin more than Molly made; and Milly, Molly's mother, Molly and Milly's mother made fifty muffins, but Milly and Molly's mother made four muffins more than Molly and Milly's mother made. So murmur now how many muffins Milly made.

31. The Trials

Teddy looked up from the sports page as his father entered the room. 'I don't understand this, Dad,' he said. His father looked at the paragraph, which described some trials down on Pendine Sands: speeds were checked over a 12-kilometre run, it said, the cars being timed at 6 km, 9 km. and at 12 km. 'What's wrong with that?' he asked.

'Well,' replied Teddy, 'it says one car averaged 210 kilometres an hour for the first 6 kilometres, 252 for the next 3 kilometres, and 315 for the final bit. I make it an average of 259, but that's not what the paper says.'

What was the average speed for the 12-km run?

32. How Green is the Garden

Spring was in the air, and Gus was in the garden marking out the new lawn.

'I don't like the shape: it's too square,' commented his wife. 'But if you make it six feet longer and three feet narrower, there'll be room for a pathway up the side. Besides, as the area will be the same, you won't need extra seed.'

But her mother had other ideas. 'Who wants a path?' she asked in that tone which made Gus squirm. 'Make it eight feet shorter and six feet wider and you'll still get the same area of lawn and there'll be space for vegetables at the far end.'

There was no arguing against that, so now Gus spends his evenings hunting slugs and caterpillars. But what was the area of the lawn?

33. Time for Church

'Ding!' pealed the bell of St Mary's, its silvery tones ringing high and clear. 'Dong!' boomed the deep throaty voice of St Mark's in reply, but just seven seconds later. And then they were silent.

Kit, relaxing lazily after a hearty Sunday breakfast, reached for a sheet of paper and began to scribble; ever since the bells had started simultaneously less than an hour previously, he had been amusing himself keeping count of their peals. They had tolled at regular but different intervals: one exactly a hundred and seven times and the other sixty-eight times.

'Come along, dear,' called Kate from the hall, 'we mustn't be late again.' 'Just a moment while I finish this,' replied her husband, not deigning to look up. But Kate wasn't standing for any such nonsense. 'You and your problems! You can finish it after church,' she told him, rushing into the room and grabbing the paper from his hand.

So Kit donned his hat and accompanied his wife dutifully. He hadn't actually timed those bells, but somebody had told him that each pealed at intervals of an exact number of seconds. He couldn't help thinking about that timing, especially during the sermon; but it wasn't until he was back home that he worked out the respective intervals of the bells.

What do you make them?

34. Betty's New Frock

It was late on Saturday afternoon. Mike jumped up from his chair as his wife came in. 'Let's see the new frock, dear,' he said, helping her out of her coat. 'But I didn't get it,' she told him, 'and now it's too late as they'll have closed.'

'You went specially for it! What happened?' Mike was puzzled, so Betty explained: 'I went into Jenkins on the way and got some nylons, and then I hadn't enough for the frock.'

'But they know you well enough: surely you could have paid on Monday?'

Betty shook her head. 'No, and it was a new girl: I offered to pay nine-tenths down, just what I had in my bag, and the rest next week. But they wouldn't do it: the girl said the sale would be over and it would be fifteen per cent more on Monday, and they'd have no difficulty in selling it.'

There aren't many husbands like Mike. First thing Monday morning he was at the shop, and bought the frock. The price had gone up as expected, and he had to pay £7·15 more than the deposit his wife had offered; but that didn't worry him.

What did the frock cost him?

35. A Wild Look in his Eyes

'For Heaven's sake! What's come over you today?' cried Mrs Green, dropping the bottle of cream into her basket, 'I only asked the price.'

'Yes, Ma; and I'll tell you again,' the man replied, a wild look in his eyes, 'it's a quarter the price of a quart, and a quart and a quarter would cost you a quarter of a pound and a quarter of a quarter of a pound more than a quarter the price of a quarter and quart.'

A couple of uniformed men rushed into the shop at that moment, and hustled the poor man out into a waiting ambulance. But what was the price of that half-pint of cream?

36. The Boss wasn't Amused

Jim's a crank, one of those peculiar people who enjoy problems and crossword puzzles! And the worst of it is that he always expects others to join in the fun.

Today, when his new Boss asked him his age, Jim answered: 'If all you had were just two or three times as many pounds as I would have ten-pences if all I had made twice as many five-pences as my last boss would have if all he had made three times as many pennies as my age in years, you would have three pounds more than twice as many pounds as I was years old five years ago.'

It's his birthday today, but tomorrow he'll be looking for a new job. How old is Jim?

37. A Wise Wife

'Listen, my dear,' remarked Mr Jones one evening as they sat in the living-room after supper. 'This is really interesting.'

Mrs Jones could never understand her husband's morbid interest in figures, but after four years of marriage she knew it was wise to humour him over his strange hobby. She looked up from her reading with a smile. 'Tell me, darling,' she begged.

'Well,' he told her, glancing at a piece of paper in his hand, 'I've added together the year you were born, the year we were married, and your present age. It makes a total of thirty-nine hundred, which is exactly a hundred times my age.'

Mrs Jones seemed to be suitably impressed. 'What an amazing thing!' she exclaimed.

In what year was this scene of domestic felicity enacted?

38. Miracle at the Town Hall

Vic stopped abruptly, much to the audible annoyance of the crusty old gentleman—apparently a retired army man, from his flow of language—who was walking close behind. But Vic was far too overcome with emotion to pay much attention to mere man in face of this miracle.

He looked at his watch, and then again at the great clock of the Town Hall. It was indeed true; he had not been mistaken: the clock was actually showing the correct time!

Vic hastened home to tell his good wife the astonishing news. A little later, telling the tale as calmly as he could, he recalled that the minute hand had been exactly three minute-divisions ahead of the hour hand, and that both hands had been precisely at minute-divisions.

At what time did he witness this modern miracle?

39. Payment Overdue

'Just a moment, you wait!'
Shouted Alf, quite irate,
At Doug, who'd come dunning for dough.

'If you'd invoiced us right
We'd have paid you on sight:
It's all your own fault we've been slow.

'You'll remember we bought
Just a few of one sort
At eighty-nine pence: that you know.

'As regards all the rest,
Which you said were your best
And priced, you'll admit, not too low,

'We bought less than a score
At a pound twenty-four,
For that was as high as we'd go.

'So the total should be
Nineteen pounds thirty-three.
I'll pay you right now what we owe.'

How many had they bought at each price?

40. Three Men at the Bar

They sat at the bar, sipping their beer while they discussed that party where they had first met. It was one of those 'd'you remember?' conversations, and they only disagreed about the date.

Jack said it had been Thursday, May 8th: Harry agreed on the month, but was sure it was a Tuesday and on the 10th: Geoff insisted it had been June 8th and a Friday at that.

The argument became quite heated and even involved the barmaid, who told them her birthday was on April 1st, and it had been on a Tuesday that year.

In fact, none of the men was completely right. But, amongst them they had mentioned the correct month and the correct day of the month and the correct day of the week. One had made only one true statement, one had been wrong on only one point and one had been completely wrong.

Assuming the barmaid was right, what was the date of that party?

41. Buying Christmas Cards

Having been dragged into the crowded shop by their wives to help choose Christmas cards, Bill and Bob were very glad to complete their purchases and escape to a nearby pub for refreshment.

Sitting over their beer, the four started comparing their cards. It turned out that each couple had bought twenty: Bill had bought three more than Anne, and Elsie had only bought four.

This may seem irrelevant, but which of the girls was married to Bob?

42. Sunday in the Warehouse

Scuttling around the warehouse on Sunday morning, Ron met his old friend Roy, hurrying towards him. 'No time for gossip now,' called Roy, waggling a shiny wing-case as he passed, 'we'll meet the other side.'

The two cockroaches continued on their respective ways, close up to the foot of the wall. But Ron turned back some moments later: 'I'm hungry,' he thought; 'I'll get back home.' After retracing his steps a matter of only two metres, however, he changed his mind about this: Roy would be worried if he didn't turn up on the other side as expected. So Ron turned once more, and continued his progress in the original direction.

Perhaps you don't know that old warehouse down on Commercial Road? It's just fifty-three metres around the edge on the floor, and Ron is the champion sprinter of his pestilential little community. Roy isn't slow but, at the normal speeds, he does only eight metres to Ron's eleven. On this occasion they both scuttled along at their normal speeds from the moment they first passed each other to the time they met again on the other side of the place: and Ron wasted no time at either of his 'turns'.

So how far had Roy travelled when he did meet Ron at last on the other side of the warehouse?

43. Jill got her Sundae

'How much have you got?' asked Jill, as she followed her brother into the sweet-shop.

Jack fumbled in his pockets. 'I've only got ten-pence, five-pence, and two-pence pieces; but fifteen altogether,' he replied. 'Yes, that's all,' he went on, after checking his cash again, 'and if the five-pences were two-pences there'd be four times as many two-pence as ten-pence pieces, but if the two-pences were ten-pences I'd have twice as many ten-pences as five-pences.'

He jingled the coins, laughing at the expression on Jill's face: 'And I'll buy you a special sundae if you guess how much I've got.' And how much was it?

44. Winter's End

Popular as alphametics are, you may still not be familiar with this type of puzzle. Here is a really easy one.

Each distinct letter in the addition stands for a particular but different digit. Only a thorough knowledge of basic elementary arithmetic, coupled with sound logical reasoning, will be required to discover what SNOW represents.

```
  W O W
  W O W
    N O
-------
S N O W
```

45. Five Threes and More

Here we have a regular but not too simple division-sum. Five of the threes are shown, and the other figures are represented by crosses; there is at least one '5' in the calculation. It all looks far more difficult than it really is, so now you are asked to find the 5-figure number.

```
x x 3 ) x x x x x ( x x x
        3 x x
        -----
          x x x
          x 3 x
          -----
            x 3 x
            x x 3
            -----
            - - -
```

46. Mike's Morning Walk

For the past few weeks Mike has been going to the office by bus: it stops just by the end of his road and is really very convenient. His wife is delighted to have the use of the car, and is quite boastful about his generosity: only to himself does Mike admit that parking troubles in the City had anything to do with it.

Striding along the pavement today, Mike noticed that he takes two of his normal paces to each of the big concrete slabs. When he lengthened his pace by six inches he took five strides to three slabs.

So now can you work out the length of his normal pace?

47. Not a Square Deal

Mr Flatt groaned when the old man walked into his office: he'd have to work for his money this time. For this was Mr Kirby, eccentric but very rich. And he came straight to the point, telling the estate-agent that he wanted to buy a rectangular plot which must comply with certain explicit conditions.

'Four times the frontage, added to three times the depth, must come to exactly a hundred and twenty yards;' he said, 'and each must be an exact number of yards and of course I want the largest possible area.' And with that he stalked out of the office, leaving Mr Flatt with no mean problem.

But the plot was found and the price was pleasingly high, and Mr Flatt wouldn't mind a few more like that. But what do you make its area?

48. He gets it from School

They've started a craze for problems at school, and now one rarely gets a straight answer out of Larry on any question at all. Today his mother asked the boy how much he had spent for her at the grocer, and his reply was a good example of the way he is these days.

'If the bill had shown half as many pounds and twice as many pence,' said Larry, in seeming seriousness, 'I would have spent £2·58 less than I'd have spent if it had shown twice as many pounds and half as many pence.'

Now, what could the poor woman make of that?

49. Jill in the Pet-shop

It was a very small and smelly pet-shop, but Jill was quite fascinated. Her parents having moved into a new block of flats where animals were banned and children only just tolerated, she was thrilled to have found this little haven so close to her new home.

The birds, somewhat bedraggled in their cages, didn't interest her so much; but she loved the puppies and kittens. The proprietor had been watching her hopefully: 'Perhaps your Dad will buy you one,' he ventured at last.

'No, it's not allowed at home;' replied Jill, 'but may I come in to see them sometimes, please?' She smiled, adding, 'I'll bring food for them if you tell me how many you have.'

The old man loved children almost as much as animals. 'Come in as often as you like, young lady,' he told her, 'but don't bother about food. I have eleven more animals than birds; but if I had as many animals as I have birds and as many birds as I have animals, there'd be only four-fifths as many legs altogether as there are now.'

Jill thought that sounded funny, and it stuck in her mind. Later that day she told Jack about the shop, and repeated the old man's words: and Jack worked it out for her.

How many birds were there in the pet-shop?

50. Some Red, Some Blue

Simple Simon has been working for some days in that office down Church St, running messages and generally making himself useful.

Two days ago he was sent out to get some red and blue pencils, ten of one and six of the other: for these he showed an expenditure of £1·10 in the little record they had told him to keep. Yesterday he went to the shop again, and for the same pencils, but five of each this time, and he charged the firm 50p as the cost of those. And today, on returning from the shop with a third lot of the same pencils he recorded an outlay of £1·26 for six of one colour and a dozen of the other.

Unfortunately for Simon, his employer was in a bad mood when he found the boy putting away the little book after making that last entry. 'One of these entries is wrong,' he growled, pointing to the records of these three days. And so poor Simon has lost his job.

Can you spot the wrong entry?

51. Those were the Days

Old Gil Giles is the biggest farmer around where I live, and he never tires of telling how he started. 'I was a labourer over with Sam Miller's dad until after I married Mrs Giles, ten shillings a week and a cottage,' he'll say, chewing the end of a fat cigar. 'Then I rented a little place and used our savings to get some stock.'

He generally pauses here, to make sure he has your attention. And then he'll go on: 'Things were very different in those days. I bought pigs at nine shillings, geese at tenpence and chickens at sixpence: a hundred head altogether and they cost me just five pounds.'

Those were the days of twelve pence to the shilling and twenty shillings to the pound. But you can work out how many geese he bought?

52. Cheaper for Two

'Sorry I've been so long, dear,' said Ruth, as she came out of the ironmonger's and got into the car. Her husband is a patient man, but sometimes unexpectedly observant. 'That's all right,' he told her, 'but I thought you only wanted a kettle.' He put her parcels on the back seat: 'Why buy a saucepan too, when we've so many at home?'

'Oh, but it was such a bargain:' explained Ruth, 'if I'd bought them separately five-ninths of the total cost would have been on the kettle, and I would have had to pay £1·30 more altogether.' She flashed him one of her most winning smiles. 'That's what the man said, so you see he really let me have the saucepan for £2·10.'

You can't argue against that sort of thing, and of course a wise man's wife is always right. But what would the kettle have cost by itself?

53. How Happy would I be with Either

'Come along, Garry, make up your mind!' cried his mother impatiently as the boy stood there eyeing the candies. Garry sighed. It was always like that when he took his mother shopping.

He only had a few coins to spend, and it was really very difficult to make a quick decision on such an important matter. 'I've got just enough for four ounces of those nut creams,' he said, 'but the chocolate fudge is cheaper and I could just buy twelve ounces of it.'

His mother was coldly practical. 'Get both,' she told him, 'the same quantity of each, and don't waste any more time.'

And that is what he did, spending all his money on the purchase.

How many ounces did he buy altogether?

54. The Road to Zirl

'Now what do we do?' asked Jim, stopping the car where the road ended in a deserted quarry. 'You're the navigator!'

His wife looked up from the road-map. 'We must have gone wrong at that fork an hour ago;' she told him, 'you remember it said "Zirl 12 miles" but we weren't sure which way it pointed.'

Jim leaned back and lit a cigarette. 'I suppose you're right again,' he laughed, 'but we've been driving four hours and I make that eighteen miles more than we should have done.' He started turning the car: 'Anyhow, it's your responsibility.'

They reached Zirl without further difficulties, arriving not too late for an excellent little dinner in the quaint old hotel.

Assuming that Jim drove at a steady speed the whole time, how far is Zirl from where they had started out that day?

55. It was Christmas

'There they go,' growled Harry, looking across the room to where his wife was fiddling with the radio, 'and now we'll have no peace.' Anne smiled: it was the same every Christmas; the same grumbles, but she knew very well he would let no child ring their bell in vain.

'I'll do it,' she said, picking up her bag and going out into the hall. She returned a moment later, however. 'Give me some change, darling;' she asked, 'if I gave them each twopence I'd have a penny over. So give me fivepence and then I can let them have threepence each.'

Harry didn't ask how many children there were singing outside the door, for he'd worked it out almost before his wife had finished speaking. But what do you say?

56. The Missing Miss

They were arguing about those two very pretty girls who had been at the party last Christmas. John had referred to them as Mary and Jane, and that started the argument. 'You've got it wrong,' said Jim, 'their names were Elsie and Gwen.'

Joe agreed with Jim as regards Elsie, but he thought the other girl was Anne. Jack was in no doubt about the names: 'I saw far more of them than you did,' he boasted, 'and of course they were Anne and Mary.'

Between them, the four boys certainly did mention the correct names. Jack gave one name right but the other wrong, and so did two of the other three. The fourth boy was completely wrong, however.

Now, the strange thing about all this is that there had been no girl at the party with one of the names mentioned. What name would you say that was?

57. A Hopeless Case

Professor Brayne is so absent-minded: they say he was seen one wintry morning, strolling down The High with thick wool mittens on his feet, carrying his shoes.

In a car-park, on one occasion, he couldn't find his car: he'd forgotten the number and even the make. 'It's four figures,' he told the man, 'and the first is two more than the last. And they are all different and there isn't a zero.'

The Professor paused, but then remembered something else. 'You'll find the second figure is more than the third, and it's all the exact reverse of my telephone number.'

'But what's that, Sir?' asked the bewildered attendant. 'Unfortunately,' replied the old gentleman, 'I've forgotten that too, but I remember noticing that the two numbers add up to 16,456.'

'Well, if you can remember all that,' but he didn't finish the sentence: the attendant didn't want to hurt anybody's feelings. But he worked it out correctly, and so can you.

58. Fair's Fair

Uncle Frank had sent Jack a lot of foreign stamps. Now the boy had them all out on the rug and was sorting the stamps with his sister's help. Two of them were badly torn, so they were thrown away, and then Jill claimed her share.

'He didn't say how many were for you,' Jack told her, 'but I'll give you a number which, multiplied by itself, makes half of all he sent us.'

'That's big of you,' cried Jill, grabbing a good handful of stamps and making for the door, 'but it would still leave you just eight-ninths of what he sent, you greedy beast!'

So how many stamps did Uncle Frank send?

59. Just a Simple Routine

If you follow my rule with fifteen,
Then you'll surely get seventy-four.
The routine can then also be seen
With what number you wish—say a score.

With a score you get seventy-one,
But with five just an eighty you'll see.
So perhaps you can figure for fun
What the answer with thirty will be.

60. A Man and his Sorrow

'You pay me, Madam,' remarked the assistant, handing her the parcel. Mrs Ryle gave the young man four crisp £5 notes, and glanced at the bill in her hand. But something was very wrong. 'I think you've made a mistake,' she exclaimed, 'and you've charged me one pound sixty-two too much.'

The assistant sighed: it was so very difficult to concentrate when every thought was of his poor Beryl, and of the lonely years ahead. 'I'm terribly sorry,' he mumbled humbly, looking at the bill, 'but I seem to have written down the pence as pounds and the pounds as pence.'

Mrs Ryle could see the sadness in his eyes, but what could she do? 'You also reversed the order of the figures in the pounds when you wrote them down as pence,' she told him. 'But we'll get it right now.' She spoke kindly, for she had also known unhappiness.

How much was the correct change out of her notes?

61. Three Men and their Smokes

Jack accompanied his friends into the tobacconist's. Seeing his favourite brand on the shelf, he bought some cigars at 25p each. Jim bought a cheaper brand at 20p, and Joe chose cheroots at only 15p.

One of the three bought nine 'smokes', while the other two bought five each. Between them they spent an exact number of pounds.

Can you see which of them bought nine?

62. A Day on the River

It's a pleasure to take a trip on the river in the little ferry which runs up and down between Tormill and Torbridge. Stretched out in a deck chair, watching the low green banks slip by, Hank felt at peace with the world. He had bought a return ticket and was now on the way back home to Torbridge.

The run up-river had taken just one-and-a-half hours, and he had been told the return trip would take forty minutes 'if all went well'. Lazing there half asleep, he found the steady beat of the engine most soothing. 'Chug, chug, chug,' it never varied; the same smooth rhythm all the way up to Tormill that morning and now still the same.

But suddenly there was silence, a hush which brought Hank out of his chair. Calling up to the tiny wheel-house in some alarm, he asked what was amiss. 'Engine's dead,' was the laconic reply, 'but we've only three miles to go and plenty of water. We'll drift it in thirty-six minutes.' And so it was.

How far is it from Torbridge to Tormill by river?

63. A Man came Back

It was early in May 1956, and Beryl and Bob were spending a quiet Sunday afternoon sitting in the garden. Close at hand their children sat reading; Garry and Doreen, three-quarters her brother's age.

'It's strange to think back to this day just eleven years ago:' mused Bob, 'I was over there in Bremerhaven, probably sitting on a heap of rubble, far, far from home.'

Beryl took his hand and squeezed it. 'I was here in this lovely garden,' she whispered, 'counting the days to your return. And even Doreen, only half Garry's age, seemed to sense my relief and happiness.'

In what year was Doreen born?

64. Horror in the Night

I was crouched in one corner of that vast bare hall, alone with my fear—but not alone, for all about me were the voices, whispers full of menace, racking my very soul with their strangely familiar questions. 'What did Mary play?' they asked, and 'How spake they in Kalota?' and 'When did Ken return home?' There was no respite. The air was filled with questions. I must answer, but I knew not the answers to those chorused and peremptory demands. It was horrible.

In that great hall, its floor and ceiling a dull, dead black, its longer walls a deep blood-red, its shorter end-walls a sickly yellow, my mind knew only one number: ninety feet; the length of the chamber was branded in my fear-crazed brain. The width and height of the place were less, but that was all I knew.

Then the voices were no longer there. Looking up, close into the most distant corner of that hall, I saw a Thing: a monstrous many-legged horror bathed in pulsing green light. 'You have just one last chance, you puerile problemist,' it thundered. 'If I crawled to you now, creeping ever nearer by the shortest possible route to tear you limb from limb, which colours would I traverse?'

I froze, speechless with terror. But the Thing continued: 'Figure it out if you would live.' Those writhing legs began to show a purpose in their weavings. I screamed, and found myself awake, still screaming. But what was the answer?

65. The New Flat

'What a long road! It sounds horrid,' objected Helen, when her husband showed her the advertisement of that flat in Clarendon St. 'But it seems just what we're looking for,' replied John, 'and the rent's quite reasonable.'

They had just arrived from Chester and had taken a room at the Midland Hotel with the idea of moving as soon as they found what they wanted. And Helen did want to be fairly central. 'Just look at the number, well up in the hundreds;' she insisted, pointing to the paper, 'that must be miles out.'

But John laughed. 'Not so much temptation to spend money,' he said, and went on to stress the apparent advantages of that flat as compared with others which were advertised.

His wife, however, was quite prepared to take his word without a lot of talk: she doodled aimlessly on the margin of the paper, paying little attention to his words. Suddenly she interrupted him. 'That's funny,' she cried, indicating her scribbling; 'if you shift the last figure of that house and put it in front of the first figure, and then add that new number to the original, you get one one eight one: that's your old home number followed by mine!'

This was an omen not to be disregarded: they went right to the place, and liked it so much that John rented it there and then. But what do you make the number of the house?

66. The Luncheon Party

'We've all remarked on it many times before,' observed Mr Bankes, 'but it still gets me. Where else in the world would you find a lawyer, an author, a dentist and a banker at one table, and bearing the names of Law, Penn, Bankes and Tooth?'

'It's worth sending in to Ripley,' replied the dentist, whose surname corresponds to Mr Penn's profession, 'especially as our names don't agree with our respective occupations.'

It was indeed extraordinary. But what do you make Mr Tooth's occupation?

67. It Fell from the Sky

They still can't agree as to which of them saw it first. But the object was certainly warm when they found it. The boys claim it came from one of those flying saucers, and Jim even swears he saw the alien ship swoop down and up with a flash of flame before that lump of calcined rock crashed into the barn behind the farm. Of course it could have been a plain meteor, but why be so mundane?

Jack says it weighed seventeen ounces, but Joe's guess is twenty-six. John, who says he's worked it out from what he calls 'terminal velocity', insists it must have weighed twenty-one ounces. Jim's estimate is one ounce less than John's.

In fact, one of those estimates was only two ounces out; two of the other three were exactly equal in the amount by which they differed from the correct weight.

Your guess is as good as mine, so what do you say?

68. A Doubly-true Alphametic

In this addition each distinct letter stands for a particular but different digit, as is the usual way in alphametics. But the words themselves add up correctly, hence the title.

Start by deciding what digits must be represented by N and by E. That only entails a little sound reasoning. Then what about the letters I and O? You should be able to continue to the unique evaluation of SIXTY.

```
  F O R T Y
      T E N
      T E N
  ---------
S I X T Y
```

69. Those Handy Cartons

Tam, Tim and Tom were sent out to buy the weekend supply of milk. They were told to get seven and a half litres. On their return, the boys trooped into the kitchen and stacked an imposing array of litre and half-litre cartons on the table.

'But why so many half-litres?' asked their mother in some surprise. 'You know it costs more like that.'

'You didn't tell us,' replied Tam, 'so we each bought the same quantity of milk.'

'But we each got a different number of cartons,' said Tim. 'And I bought the most half-litres,' added Tom. How many half-litre cartons did Tom buy?

70. Tony never used to be like That

Tony used to be such a polite little boy, but now his mother says he must be picking up bad habits at school. A few days ago, for instance, he was really so cheeky when his mother introduced him to the wife of the new minister. The good lady asked Tony his age, and was quite taken aback at his reply.

'Twice my age, taken away from the square of my age,' the boy told her, 'comes to seventy-seven more than Pam's age.' Grinning impishly, he added: 'And please don't bother about vulgar fractions.'

Tony's mother was most embarrassed, but she explained that Pam is the boy's younger sister, and of course the minister's wife can forgive a lot where such a generous supporter of the chapel is concerned.

But how old did all this make Tony?

71. A Quiet Evening at Home

Harry and Tom spend many a quiet evening playing draughts. They play for cigarettes, the loser handing over one at the end of each game.

Last Tuesday they had a long session which resulted in one of the two friends winning a nett total of ten cigarettes from the other. At the finish they each had six cigarettes, although Tom had smoked three and Harry seven during the evening's play.

How many cigarettes must Tom have had when they started playing that evening?

72. So Soon Forgotten

The girls were discussing three boys they had met on their summer holidays. There was no doubt at all as to the first names, of course, but they just could not agree on the surnames of the young men.

Peggy said she had spent a lot of time with Jack Lennox, but she had not liked John Everard. 'I quite agree about John Everard,' Pam nodded, 'but surely Jack's surname was Best.'

Now Pat had her say, insisting that the Best boy was John, and the other two were Jim Lennox and Jack Everard.

As it happens, each of the girls was right about at least one of the young men: the surnames were certainly the three they mentioned. So can you link the names correctly?

73. What a Tie!

'Oh boy, oh boy!' cried John, as his friend strolled into the crowded newsroom. 'Just look at his tie!'

'Yes, I know,' rejoined Dick, not at all put out by the laughter around him, 'but my wife chose it and you know what that means.'

John nodded understandingly. 'She chose and you paid. Two pounds wasted. That's marriage!'

'Cheap at the price,' chuckled Dick, 'but my tie didn't cost that much. It actually cost twice as many pence less than £1·40 as six would have cost in pounds.'

Even John was quiet for some minutes after that! But perhaps you can work out the price of the tie.

74. A Tale of Three Monkeys

'We'll share them out in the morning, boys,' grunted Flip. 'It's too dark now.' The three monkeys were tired. After toiling all day collecting a great supply of turtle-eggs, they were only too glad to curl up on the soft, warm sand for some welcome sleep. And so all became quiet, except for the murmur of the wavelets lapping along the beach.

But soon the rising moon wakened Flip. He peered at the other two and saw that they were sleeping. Edging cautiously to the pile of eggs, he broke one, swallowed its acrid contents, and scrabbled the shell into the sand. Carefully he counted the rest; carefully he counted out for himself exactly a third of them, which he hid deep in the sand where he again lay down and slept.

Time passed. Then Flap opened his eyes and saw the others fast asleep. He was hungry. Swallowing the slimy mess of one addled egg, he hid away just a third of the remainder. Then, well content, he fell asleep again in his place.

Soon there was a gentle stir and a rustle as Flop opened his eyes. Seeing the others deep in slumber, he gulped down an egg and then took for himself a third of the much reduced heap. These he buried in the sand, and within minutes he was dreaming again.

And so when they wakened with the dawn, they shared the eggs which still remained, each taking an exact third and hoping his duplicity would not be suspected by the others. We don't know how they dealt with those hidden eggs, but we do know the three must have concealed between seventy and eighty eggs between them. Perhaps you can work out the exact number.

75. A Game of Canasta

'Let's not play point-stakes today,' said Mary, as the three friends sat down to play Canasta. 'I suggest the loser should double whatever money each of the other two has at the end of each game.'

Her two guests agreed with this novel idea—after all, Mary was hostess—and so they settled down to business.

Gwen's luck was right out at first: she barely managed to meld, and she lost the first game by a disgraceful margin. She made up for it in the next game, however, which she won.

Mary lost the second game. But the third was really exciting. It developed into a hard tussle between Gwen and her hostess, each drawing lots of Wild Cards and going out several times on Concealed Hands.

After that third game, in which Clare had scored hardly any points at all, Mary proposed a break for tea. Pouring for her guests some minutes later, she remarked that she had won a little. 'I've got exactly four pounds now,' she said. The other two checked their cash also, and found they each had four pounds too.

It was a strange coincidence. But how much did Gwen have when they started to play?

76. On the Broad Highway

There was little traffic that day on the road between Glaur and Polmood. But that was the day Don drove from Polmood to Glaur, starting off at exactly the same time as his friend Jock left Glaur in his car headed for Polmood.

They both drove at their favourite speeds, each maintaining a steady speed on the almost deserted highway. When they passed each other, Jock had driven nine miles more than

Don and required sixty-four minutes to complete his run to Polmood. Don reached Glaur thirty-six minutes after Jock reached his destination.

What do you make Don's speed that day?

77. One Rainy Day

It was really too bad that it should rain today, of all days. The children had been so much looking forward to their big picnic, and now it was out of the question. The previous night they had gone to bed excited and with great hopes, for the forecast had been 'fine and warm'; then this morning they could only see rain and clouds.

But they had all assembled at the Browns' house as arranged: the Lamberts from across the river, the Beale boy from next door, and the Harvey kids from up the hill. And now the children were milling around in the dining-room, bored and disappointed, not knowing what to do with themselves.

Then Uncle Frank came to the rescue. Striding into the room, he slapped a handful of ten-pence, five-pence and one-penny pieces on to the table. 'Here you are, kids,' he told them. 'Divide this equally among you; there are forty-five coins, but you'll have to arrange change out of your pockets.'

Sharing out the money, which came to £4·07 altogether, kept them busy for quite a long time. Each received more than twenty pence. So how many children were there, and how many tenpences did he put on the table?

78. Humpty Dumpty

Poor stumpty Humpty Dumpty! So loyal was he, so determined to see every detail of the great Royal Pageant. And now? 'All the Queen's horses and all the Queen's men—' well, you know how it goes.

Had he only lived three years longer, and been born (or is it 'laid'?) one third of his age at the time of his demise earlier, he would have died when he was as old as he would have been if he had lived to the age of five years less than twice the age at which he died.

How old was he when he had his fatal fall?

79. The Spreading Chestnut Tree

'That's a fine chestnut,' commented Ben, admiring the magnificent tree which towered in front of the farm-house. 'Yes,' replied the farmer. 'My Dad planted it when he built this house. He used to tell me the sapling was eight feet tall when he put it in, and he said it grew exactly the same amount each year.'

'But d'you mean every year from the start?' asked Ben, who always liked to get his facts right.

'No, I wouldn't go as far as that,' admitted the old farmer, 'but it was true for the first dozen or so years, and I remember being told that it was one-twelfth taller at the end of the ninth year than a year before.'

How tall must that tree have been at the end of the tenth year?

80. Only Snails

The two snails were hurrying along on their lawful occasions on the cylindrical column three metres around. Sammy was one metre up from the base at the moment when Simmy, on exactly the opposite side, was three metres up.

Just then some molluscan sixth sense made each aware of the other's proximity. Spring was in the air, and even snails have their moments.

'Which way will he come?' wondered Simmy, halting in her slimy slither. It was a long wait, but Sammy didn't disappoint her.

How far did he travel to her by the shortest route?

81. A Tale of the Cats

The lucky cats in Stratton Street
Had seven mice apiece to eat.
The rest made do
With only two:
The total score
Being twenty-four.
How many cats ate mousie meat?

82. A King was Chosen

And now were the Elders assembled together in the Temple; and the sound of their voices was like unto the noise of thunder as each spake unto the other asking: 'Whom then shall we choose to be our King to rule over all Kalota?' And in that multitude there were many who would choose Kar the son of Knok; and there were some who called for Kapel: for it had been so ordained, and no other could be chosen, only of these two who now stood before the Elders.

Then came Keris the High Priest and spake, saying: 'All ye who choose Kar the son of Knok raise ye now the right hand.' And straightway were their right hands raised and they were like unto a field of corn before the reaping; and Keris went among the Elders and he counted; and two score and one of the Elders had not raised a hand; and eight score and five had raised only the right hand as Keris had so adjured them; and some, seeking advantage for their choosing, had raised two hands; and all the hands that were raised at that time numbered three more than the whole number of the Elders in that congregation.

And the High Priest went down from the high place and laid his hands upon Kar and blessed him and turned to the Elders, saying: 'Hear ye now all ye Elders of Kalota that are gathered together in this place from the farthest ends of the land, and let it be known by all men that ye have chosen Kar to be your King.'

That's what it says in the Chronicle of Kalota, but it would be interesting to know how many Elders voted for Kar.

83. Simon goes Shopping

Simple Simon was sent out to get some handkerchiefs for himself. He came back with three, and he had also bought half-a-dozen for his sister Sarah. When his mother asked how much he had paid, however, the boy wasn't sure.

'I know mine cost seven pence each more than Sarah's,' he said, 'and I noticed that what I paid for hers was the same as I paid for mine but the other way round.'

His mother couldn't quite follow this. 'What on earth does that mean?' she laughed.

It was so clear to Simon. 'Well, the two amounts were the same two figures but in different orders, like 53 and 35.'

Perhaps you can see how much Simon paid altogether.

84. Top Secret

The boys knew their father had been doing something rather mysterious during the war. He always called it 'Intelligence', but then you'd hardly associate that with his having been so long at the Admiralty. Now he had just finished telling them about ciphers and codes, and was back to his newspaper.

But a seed had been sown. It wasn't long before Jack came over with a scrap of paper and demanded attention. 'Add ode to ado to make deed,' declaimed the boy. His father pondered a few moments, making some notes on his paper, and then he nodded. 'That's a queer code if you like,' he chuckled, 'using letters for figures.'

What was the little addition sum?

85. A Profitable Round

'What can I get you?' asked the grocer, seeing this well-known customer wandering rather aimlessly around the shop. 'That depends!' was John's surprising reply. 'I'll buy two pounds' worth if you'll first double what little money I have in my pocket.'

John is quite a character in the village, and always a good spender. The grocer decided to take a chance, and so the transaction was completed, and on a cash basis.

Leaving the shop some time later, highly amused by all this, John felt tempted to try the same tactics in two other shops where they value his custom. As expected, the owners of both were very ready to humour the whim of such a good customer even at the risk of some slight loss.

When John finally left the third shop he found he had spent all his money and had not even a penny in his pocket, but his car was stacked with a varied assortment of goods. How much did he have when he entered the first shop?

86. No Head for Figures

Professor Brayne was in trouble again today: it's hard to understand how such a brilliant mathematician can have so poor a head for simple figures. There, in the crowded car-park, surrounded by cars which all seemed to look like his, he had to confess that he had quite forgotten his number!

'Isn't there anything you remember about it?' asked the attendant.

'Well, yes,' replied the old gentleman. He smiled, quite unembarrassed by the whole ridiculous situation. 'My boy once told me that, if I shifted the first figure and put it after the last, I would get a new four-figure number one more than three times the original.'

The attendant pulled out a scrap of paper, and found that number while the Professor went on muttering to himself.

What would you make it?

87. In Kalota there was a King

One hears more and more of the strange customs which make Kalota so interesting to visitors. But perhaps the strangest of all is the convention about speaking the truth.

The men never speak the truth, but never! And that really makes it very easy to know what they mean in conversation.

The women, on the other hand, always alternate truth and falsehood or vice versa, statement by statement. That does make things difficult but, compared with what happens in other countries, there's quite a lot to be said for this feminine vagary: a Kalotan woman does speak the truth half the time.

Now three of these delightful people were talking about their beloved King Kormon, who died some years ago: there seemed to be some doubt as to the year of his death.

'It was the year of Karan,' declared one of the three. 'You're wrong,' said the second Kalotan, 'it was the year of Kelkis.' The third Kalotan appeared to agree: 'Definitely the year of Kelkis.'

As a matter of fact their King Kormon did die in one of those two years, but unfortunately, I've forgotten which. And, as I've also forgotten the sexes of those three Kalotans, I'm still trying to work out the truth. But what do you make of it?

88. A Monkey on the Job

Today I met a monkey, but this was down High St. where they are building the new flats. He was dangling at the end of a rope, holding on with one paw and scratching happily with the other. Going closer, I saw that the rope passed over a pulley with a weight at the other end, and the two ends were exactly level.

This was all so strange that I made enquiries from a man who seemed to be working there. Thus I learned that the monkey's age and the age of its mother total seven years; furthermore, the monkey weighs as many kilos as its mother is years old.

Intrigued by these revelations, I heard that the monkey's mother is one third again as old as the monkey would be if the monkey's mother were half as old as the monkey will be when the monkey is three times as old as the monkey's mother was when the monkey's mother was three times as old as the monkey was then.

This news staggered me, but more was to come: the weight of the weight and the rope came to twice the difference between the weight of the weight and the weight of the weight and the monkey!

I could stand no more. 'Tell me the length of the rope,' I begged, 'and let me go.' My informant shook his head sadly: 'That we don't know,' he said, 'but four feet of it weigh one kilo.

I went on my way wondering, but still seek the answer to my question.

89. Mike's Pike

'What lovely fish!' exclaimed Mike's wife, admiring the three fine pike he had brought home. 'What do they weigh?'

'The largest weighs as much as the other two,' replied Mike, 'the smallest weighs three kilos less than half the other two together, and the lot come to eighteen kilos.'

So what were the weights?

90. More Files for the Office

'Will that be all?' asked the assistant. Joan toyed with the stapler she had chosen. 'Oh, no. Mr Brent gave me five pounds. This is £2·84, and he told me to buy it first. Then I must spend the rest on files, as near as possible to two dozen quarto size and the rest foolscap.'

The man reached up to a shelf behind the counter. 'I know what you always have: foolscap at seven pence and quarto at five pence. But how many of each did you say?'

Joan was silent a moment, scribbling on a scrap of paper. 'You'll have to help me with this,' she said at last. 'He's funny about money: hates having change handed back and says it upsets his books. So I must do what he said.'

'Certainly sounds crackers to me,' laughed the assistant, trying to read her figures upside down, 'but anything to oblige a customer.' Needless to say, they did solve that problem and strictly in accordance with Mr Brent's instructions. But what do you make of it?

91. The Flatterer

'Good morning, and how are you today?' Sally greeted her neighbour as they met in the store.

'Not too good,' replied Amelia in resounding tones which belied her words. 'If you were as old as I feel, and if I had been as young ten years ago as you look today, you'd be twenty years older than me.'

Before Sally could quite determine whether this was unexpected flattery or much the reverse, Amelia went on: 'And really, my dear, you look twenty years younger than I am.'

So how much older did Amelia suggest she felt that day?

92. That Jack Again

'How old's Aunt Emily?' asked Jack, looking at the snapshots which had come in the mail.

'You can work it out yourself,' his father told him. 'She's thirteen years older than I was when Aunt Claire was half my present age. And she is five years older than Aunt Claire, and two years younger than me.'

Maybe Jack knew Aunt Claire's age, but anyway you can figure out how old Aunt Emily was.

93. She Married them Off

'What happened to those nice girls who used to run around with your Susan?' asked Mrs Jones who hadn't seen her friend for some years. Mrs Brown smiled: 'You mean Sally and Doris? They got married very soon after Sue.'

'Now isn't that nice. Such a relief for their mothers.' Mrs Jones was thinking of her own dowdy daughter who seemed destined for other things.

'Yes,' Mrs Brown went on, 'they married three old friends: Tim Morton, John Devlin, and Len Owens. Maybe you'll remember the boys.'

This was a good start for a real gossip, during the course of which Mrs Jones learned that Sal must be twenty-three; that Mrs Owens is one year older than Mrs Morton, and that Sue is two years younger than Doris. It also transpired that Mrs Devlin was as old as her husband was when he was three years younger than Mr Owens was when he was twice as old as his wife was when she was three years younger than Mr Morton was when he was half as old as his wife will be in seven years' time.

Mrs Jones came away slightly dazed, but perhaps you could tell her Susan's age.

94. No Union Rates

Bill and Jerry were taken on to dig a shallow trench from Farmer Giles's house down to the pond. It was to be 160 metres long, and the farmer told them to start at opposite ends and work towards each other: that way he felt they wouldn't waste their time.

Bill, who started up by the house where the ground was hard, was paid 30p a metre more than Jerry. Jerry was quite satisfied, as the ground was soft down by the pond and he would be able to do far more than Bill.

When the job was finished, however, each received £36·40. So how many metres did Bill dig?

95. Best Irish Linen

Sal was showing Sue the present she had bought during the lunch-hour. 'I got these for my boy friend,' she boasted, opening a box of handkerchiefs. 'They're real Irish linen, but amazingly cheap.'

'How lovely!' exclaimed Sue, noting the depth of that box. 'How many did you get him?'

Sal jotted down something on the back of an envelope before replying. 'I told you they were a bargain,' she smiled, 'and I bought one less than the price of each in pence and spent three pounds forty-three altogether.'

Sue wasn't going to waste time on that! She had only been trying to show polite interest. But what do you make the price of each handkerchief?

96. Venice or Vladivostok?

The students at Exville High School are surprisingly bright. A recent examination paper included two tricky questions which thirty-seven of those students answered correctly.

The questions were: 'Which is the more northerly, Venice or Vladivostok?' and 'Which is later, sunset at the Pacific or sunset at the Atlantic end of the Panama Canal?'

Of course some of them just didn't know. A third of the students were wrong on the latitude question, but only a quarter of them failed on the question of sunset. A fifth of the students answered both questions incorrectly.

How many students were asked these questions?

97. How Old is Jack?

If Jack were two years younger than Jill would be if Jill were two years older than half as old as Jack would be if Jack were two years younger than twice as old as Jill would be if Jill were twice as old as Jack is, he would be ten years older than he is now.

All of this makes Jack's age quite clear—or does it?

98. Tony's Birthday

'What day is Tony's birthday next month?' asked Brian, one evening last June. Beryl smiled, happy to see he hadn't quite forgotten his 'in-laws' in New Zealand. She probably didn't realise this was only one of the many little ways in which he strove to please her.

'You like problems,' she replied, 'so I'll tell you the way Tony told me years ago. Add together all the dates of the days which follow his birthday in the month, and that'll give you three times what you get if you do the same for the days before his birthday.'

So now what *was* that date?

99. The Fiver that Wasn't

'Bad news for you, Ted,' said Bob, as he entered his neighbour's shoe-shop. 'That fiver I changed for you yesterday was a dud! The bank just told me.'

The previous day a complete stranger had bought a pair of shoes for £3, paying with a crisp new fiver. Ted, finding himself short of change, had slipped over to his friend's shop to get the note changed, and so had been able to give his customer the £2 due to him.

'I'll be damned! I'd have sworn it was genuine, and he did sign on the back,' exclaimed Ted, more horrified by his own stupidity than by the actual loss. He then explained the circumstances while counting out five one-pound notes for his friend. 'Anyhow, *you* don't have to lose anything,' he told Bob, handing over the money.

What was the true extent of Ted's loss?

100. The Story of Kalia

Now in Kalota there was a man and his name was Kurtis: and he grieved for that he had seven daughters but no son had been born unto him. And Kurtis went unto Kerin the High Priest to seek knowledge of what he must do.

And Kerin spake unto him saying: 'Be patient and the Lord will bless thee and thy wife Kalia will bear thee a son.' And Kerin spake also of the gold which Kurtis must give as a thank-offering. And Kurtis swore that he would do what was required of him, and he returned to his home and did all that Kerin had enjoined upon him.

And it came to pass that Kalia bore him a son and Kurtis rejoiced and his heart was full, and he went to the Temple to give thanks and with an offering of gold.

And this was the count of his offering, even as he had sworn unto Kerin the High Priest. For each whole year of his age did he give five kalens of gold; for the first month that Kalia his wife had dwelt with him did he take back unto himself one hundredth part of a kalen of gold; for the second month that Kalia had dwelt with him did he take back two hundredth parts; for the third month three hundredth parts; and in like manner for each whole month thereafter that Kalia had dwelt with him.

And the gold which Kurtis gave unto Kerin the High Priest was one score and three kalens, for he was a rich man and thankful. And he lived in peace and happiness with his wife Kalia and with his son and his daughters, and was gathered to his fathers when he reached the age of three score and ten years.

That's what the records say, but how old was he when his son was born?

101. Draughts with a Difference

Mike and Steve had just finished their sixth or seventh game of draughts when Mary came over to the table. 'No more of that, Steve,' she said firmly. 'You can both be sociable now for a change.' Steve looked up with a grin, knowing better than to argue with his wife: 'That's all right by me, but I've won a bit and I wanted to give Mike a chance.'

Betty had come over to join them and was ruffling her husband's hair. 'Sorry, Steve,' she laughed, 'but you'll have to give him his chance another day. We're tired of sitting over there alone, whispering so as not to disturb the game.' She picked up some coins which lay on the table beside Mike. 'So tell me how much you've lost.'

'Five pounds fifty-five,' Mike confessed, 'but we played for unusual stakes. Each game we played for half of what I had in my pocket at the time, and we paid up after every game.'

He really didn't know how much he had left, for Betty had put most of it into her bag. Perhaps you can work out the amount.

102. Three Sisters

Betty, Gwen, and Jane are sisters. Betty is as much younger than one sister as she is older than the other. Gwen is seven years younger than twice the age of Jane. Jane is five years older than half the age of one of her sisters.

How old does that make Gwen?

103. Spot the Winner

Jim and Joe were arguing about last year's athletics, and in particular about the 100-yard sprint for which there had been only three runners.

'But I tell you I saw the finish,' insisted Joe, 'and Ken won the race with Steve a close second.'

But Jim shook his head. 'You may have seen it,' he told his friend, 'but you've got it all wrong. Ken was second and Larry first.'

And neither of them was right, as each had made one mistake regarding the final placing of those three runners. But perhaps you can spot the winner.

104. That was a Family

Ron is very attached to his aged grandfather, and really enjoys his stories of the past. When the old man embarks on a long rambling tale of his boyhood, however, it is sometimes difficult for Ron to keep track of all the brothers and sisters whose names are mentioned.

During one such story Ron interrupted his grandfather. 'How many brothers and sisters did you have?' he asked.

The old man went on with his tale for a moment, ignoring the question. And then he stopped, peering at the boy over the top of his glasses. 'What was that? How many of us?' There was a merry twinkle in those tired old eyes: 'You're a bright lad, so you can work it out yourself. My sister Prudence had three times as many brothers as sisters, while I had only twice as many.'

Ron did work it out. But what do you make of it?

105. John and his Bicycle

'You'll have to hurry, dear, if you want to get your bike today,' said his mother, and so John set out for the village, walking at a steady six kilometres an hour.

When he reached the cycle-shop the bike was ready and he only stayed five minutes there before starting back home. It's a good road and he cycled back at an average speed of eighteen kilometres an hour, spurred by thoughts of the supper his mother would have waiting for him.

'See, Mum,' cried John, as he entered the house, 'I've only been away five minutes over the hour.'

How far is that cycle-shop from John's home?

106. Findings Isn't Keepings

'Look, Dad!' cried Ian, rushing in with a muddy leather purse in his hand. 'See what I just found in the road!'

'You'll have to take it to the police,' his father told him. But curiosity is a very human failing, and so he added: 'What's inside?'

'Only a few coins,' replied the boy, 'pennies, two-pennies, and ten-pences, but I didn't count them.'

His father took the purse and examined its meagre contents. 'I don't expect they'll give you anything, so I'll reward you myself,' he said, 'but you'll have to work out how much there is here.' He looked again, and went on: 'The pennies and two-pennies are one more than the ten-pences; the ten-pences and pennies are three more than the two-pennies; and the two-pennies and ten-pences together five more than the pennies.'

What do you make the value of the contents?

107. Who Broke the Cups?

It's always the cups that get broken, and Mary has such an accumulation of odd saucers. So today she made a point of buying some extra cups; she chose good china, but plain white as being easier to replace.

The saucers cost 19p each but she had to pay 37p apiece for the cups, and you'd say she got really good value for her outlay of £5·58. But can you say how many saucers she bought?

108. Not for Children

'Well, dear, you must do what you think best, but it's a real problem,' said Steve to his wife as Garry came into the room.

The boy was quite accustomed to this sort of thing; so many of his parents' conversations seemed to fizzle out at his appearance. But this time he'd caught the word 'problem'.

'Here's one for you, Dad,' he said. 'When Jane was as old as the number of years Jack is older than John, she was half as old as Jack was when he was half as old as John is now. The ages of the two boys add up to eighteen years, so how old is John?'

What would you say?

109. A One, a Four, and a Nine

Take a 'one', a 'four', and a 'nine'—all three of them, but only one of each. Then, using any mathematical signs which you may require, form an expression for the number 1.

That won't be difficult, but then you have to do the same for every number in turn, working upwards as far as you can go.

As a simple example, you could show 32 as $4(9-1)$, but there aren't many as easy as that: you'll certainly run into difficulties by the time you reach the twenties!

110. An Alphametic

It may seem rather hard! But each letter in this little addition-sum stands for a different figure, and you only require common-sense and some patience to find the meaning of 'TWEED'.

```
 C A N'T
   C U T
―――――――――
 T W E E D
```

111. Sam comes Home by Taxi

'Look! Sam's just got home by taxi,' Gwen pointed across the road. 'He always comes back that way now.' Her husband nodded. 'Yes, I know. The poor fellow had his licence suspended.'

But Gwen has a horror of extravagance. 'It must cost the best part of a pound a day. Why doesn't he use the bus?' she persisted. 'Much less than that,' Fred told her, and he then went on to explain Sam's reasoning as it had been put to him only a few days before.

Living quite a distance from his job, Sam had come to the conclusion that valuable time would be wasted returning home by public transport; and then there was also his own personal comfort to consider. By bus, with one change, the journey cost only 16p; but a taxi, although charging 21p for the first mile and 3p for each fifth or part of a fifth of a mile, saved at least an hour.

Working additional overtime at the factory, and completing the necessary number of shirts at the piece-work rate of 19p, Sam reckoned he just covered the extra cost of his taxi each evening and actually got home earlier.

Gwen listened to all this, but wasn't convinced. 'What about the tip, and income tax?' she objected. Fred laughed: 'It wasn't tactful to mention that, but don't you go putting ideas into his wife's head!' Gwen wisely said no more.

What do you make the taxi fare?

112. Which Road for Knokado?

You will remember that Kalotan women never make two consecutive true or false statements: if one statement is true then the next is false, and vice versa. This is one of the peculiarities of those fascinating people, and it certainly adds to the excitement and uncertainty of life in Kalota.

Over there last year on a visit, I had an interesting experience which illustrates how this feminine vagary can be used to overcome the very difficulty it creates. I was driving to Knokado, the capital of the island, and came to a spot where the road divided: one way led right and the other left, but there was no sign or other indication as to the proper road.

A peasant woman was standing there on the roadside, so I asked her. Before doing so, however, I recalled where I was: in other countries women can never be trusted, but this was Kalota. So I asked just one question: one question worded most carefully.

The woman replied quite civilly: one word. Realising that she must surely know the way to the capital, I continued my journey in absolute confidence and so arrived safely at my destination. Later, as a matter of interest, I did find out that the other road would have taken me in quite the wrong direction.

It may amuse you to work out the wording of my question.

113. One Girl among Them

Bill, Fred and Sam are neighbours. Their surnames are Leary, Yates and Rendell, but not necessarily respectively. Each has a grown-up son, but among their six children there is only one girl.

Yates has one child less than Sam, while Fred has as many children as Sam and Bill together. Rendell has the same number of sons as Yates.

So what is Leary's first name, and how many sons has he?

114. The Telephone Number

'Is Mr Brent in?' asked Ken. 'I'm sorry, sir, but he's out and we don't expect him back today. May I tell him who called?' The telephone operator sounded quite charming and most helpful, obviously a new girl. But Ken could not wait until the morrow. 'It's Ken Keeling,' he told her. 'Can you give me a number where I'll get him?'

For a moment there was only a faint sound of rustling paper, and then she was back. 'Mr Brent did leave a message for you.' There was a suppressed giggle. 'If you double the number where he'll be and subtract your age, he say's you'll get the exact reverse of that number on the Park exchange!'

Ken swore to himself, but thanked her before ringing off. 'Curse Sam's sense of humour!' he muttered. But this was urgent, so he could only sit down and waste precious time working it out.

Of course you'll have to know that Ken is thirty-nine, and then you'll surely find that four-figure number.

115. Their Birthday

It's their birthday, you see,
The same for all three,
It's strange but it's perfectly true.

There's Bertie and Ben,
Who differ by ten;
Eight years older than one there is Sue.

Double one brother,
Plus treble the other,
Plus Sue's age makes seventy-two.

From what's on this page
You can find the girl's age.
It's really quite easy to do.

116. Simple Simon Again

Simple Simon was sent to the village shop to get some turps and return empty bottles. He returned home with five bottles of the liquid for which, he told his mother, he had paid 26p each. 'That included the bottle,' he added, 'and they allowed me their usual three pence on each of the empties.'

But when it came to handing over the change from the money he'd been given, Simon ran into trouble. He'd mixed it up with his own cash, and of course he couldn't remember how much he'd paid, and nobody knew how many empties had been returned.

'I know I gave them more than eight shillings,' said the boy, 'and I did notice the amount was exactly divisible by seven.'

What would you do with a son like that? His father, who had been listening to all this, sighed and shook his head. 'Let's hope he'll grow out of it,' he told his wife. But perhaps you can work out how much Simon spent?

117. That Old, Old Story

The old riddle starts with the host pointing to a portrait on the wall. In the modern version, which starts the same way, he might go on to say. 'Brothers and sisters have I none, but that man's father's a score and one years older than that old sun-of-a-gun, whose father of course is my father's son.' Poetically inclined, he would possibly continue: 'Forty was he when his portrait was done, two years ago by an artist for fun.'

Now, how old would the speaker be?

118. Pam's Birthday

Last week Pete and Pam had their birthday. They aren't twins, although they do look so much alike, but it just happens that Pam has the same birthday as her elder brother. And this year he noticed something rather special about their ages.

'D'you know?' he said. 'If you multiply our two ages together and then add yours and mine, you get the number of Dad's car.'

Pam considered this a moment. 'What, you get 142?' she asked. Pete nodded. 'That's what I'm telling you.'

But Pam's bad at figures, and anyhow she couldn't see that it mattered very much. Pete was right, however, so how old does that make Pam?

119. A Parcel from Each

Jack and Jill are quite good at solving the peculiar problems which result from their mother's inability to keep track of what she spends shopping. Yesterday she returned home with five parcels, having spent all her money; and she wanted to write down the details in her little black book.

'Now you can help me,' she told the children. 'I went into only five shops, and in each I remember I spent ten pence more than half of what I had in my bag when I entered.'

Jill finished first, but said nothing until Jack was ready; and they agreed on the amount. So how much had their mother spent?

120. For Betty's Birthday

'What's the trouble?' asked Mr Kemp. Joe coughed nervously: 'Well, sir, it's Betty, your secretary. We're getting something for her birthday and the boys thought you might like to come in on it.'

The manager smiled. 'Only too glad,' he said, 'and I'll give a pound more than a fifth of what you have collected already.'

'It's awfully good of you!' exclaimed Joe. He glanced at a slip of paper in his hand. 'We hadn't got two pounds yet, and that will increase the average sub by exactly forty per cent.'

How many had subscribed already when Joe saw Mr Kemp?

121. Some Christmas Shopping

'Wait out here a moment,' Mary told her husband as she turned back to re-enter the big store. 'I won't be long.' Steve was only too glad to miss another battle with the hordes of Christmas shoppers who thronged the aisles inside. 'But won't you need more cash?' he asked.

Mary looked in her bag. 'I've got plenty, dear,' she assured him. 'It's only one little thing I forgot. I only spent a quarter of my money in there, and I've still got a pound more than I would have if I'd spent a third instead of a quarter.'

That gave Steve something to think about while he waited. How much would you say she still had in her bag?

122. A Glad Goodbye

Ken was glad to see the last of Aunt Amelia. He hadn't enjoyed her visit at all. All day long it had been 'Don't do that' and 'You mustn't do this'. She was worse than Miss Prym at school.

He heaved a happy sigh as he climbed back into the car after seeing the train vanish round the bend: 'I'm sure glad the old hag's gone!'

His mother was shocked. 'Ken,' she scolded, 'don't you dare talk about my sister like that. And anyway she's younger than me.'

The boy saw he had gone too far this time. 'I'm sorry, Mum,' he told her, 'but how old is she really?'

Ken's father secretly shared his son's sentiments, but he answered the question. 'Three years ago,' he said, 'she was five times as old as the difference between Clem's and Clare's ages, and in a year's time her age will be twice their ages together.'

'But who are they?' asked Ken, never having heard of Clare or Clem. 'Just two of your many cousins, son,' chuckled his father, 'and one's a teenager.'

So how old does that make Aunt Amelia?

123. The Long Road Home

The Clarkes are punctual people. They live well out of Bristol where Mr Clarke works, and they don't envy the city-dwellers. He commutes, and returns home every evening by the same train which, 'believe it or not', always runs exactly on time. And his wife meets him at the station, leaving home at the same time every day and driving in at the same steady speed to reach the station simultaneously with the train. It's an unusually well-ordered life they lead, as you can see, but they seem very happy.

One day recently, however, things went wrong at home. Mrs Clarke is never one for excuses so we won't know the details, but she certainly left the house thirty minutes late.

On leaving the train, Mr Clarke realised that something very much out of the ordinary must have delayed his wife, and so he didn't waste time at the station but set out on foot.

Just as he was beginning to weary, his fond wife appeared, bundled him into the car, and drove back home. The old grandfather clock struck seven as they entered the house: they were twenty-two minutes late!

Despite the circumstances, Mrs Clarke drove at her customary sober speed that afternoon. So how long had her poor husband been walking when she picked him up?

124. Only Partly Right

The three boys had almost come to blows when their father looked into the room to see what all the noise was about. They told him they were arguing about their birthdays last year.

Tam contended that Tim's birthday had been a Monday and Tom's on a Friday.

'Mine wasn't Monday!' shouted Tim. 'But yours was a Tuesday.'

Tom disagreed with Tim about the Tuesday but maintained that his own birthday had been a Monday.

Their dad laughed. 'Those were the three days, sure enough, but you may be interested to know that each of you made one true statement and one untrue.' Making for the door, he concluded: 'So now you can work it out for yourselves, and let's have no more noise.'

What days were their respective birthdays?

125. Bath Night

Ron must have been in the bathroom at least half an hour. When he finally condescended to come out, his mother was waiting outside all set to give him a piece of her mind.

But the boy fore-stalled her. He explained he had been doing some school work in the bath! 'It's the tiles on the wall,' he told her. 'If each were an inch smaller there'd have to be six more tiles to the row along that wall: but if each tile were an inch bigger there'd be four less.'

His mother had little patience for such nonsense and she packed him off to bed. But how long is that wall?

126. Neighbours

Fergus and Fred lived on the 'odd-number' side of their road. Sitting in his friend's garden one day, Fergus remarked that the number of his own house, added to a hundred times the number of Fred's house, gave the four-figure number of the latter's car.

'That's funny,' replied Fred, 'but what about this? Our house numbers, multiplied together, also give the number of my car.'

That is what happened, but you will have guessed what this is all leading to. You have to know that the first figure of Fred's car number was not a nought, and then you can work out what that number was.

127. One Dud Dollar

A collector had eight Edward VII silver dollars, all identical and in mint condition. Showing them to some Canadian friends one day, he also showed them a counterfeit coin which appeared to be identical with the real dollars, but he told them that the imitation was slightly lighter than the genuine article.

Unfortunately the coins became mixed while being examined by the guests. The collector owned an accurate balance, but had no weights for it. Nevertheless, using this alone, and in only two weighings, he identified the dud dollar.

Can you see how he did it?

128. How Young is Young?

'Saw you with a good-looking young man last night at the Palais,' remarked Peg, as she joined her new friend at the bus-stop.

'Oh, that was only my brother,' replied Pam.

But Peg seemed to have been impressed. 'I'd like to meet him some time,' she suggested.

Pam laughed. 'He's much too young for you, much younger than you'd think. When he's as old as I shall be five years before I'm twice as old as he is now, he'll be just sixteen years older than I was when I was half his present age.'

Peg looked at her open-mouthed. She had never expected an answer like that. But perhaps you can work out the boy's age.

129. His Dinner Waited

A fat and foolish fly was fast asleep just one foot down from the ceiling halfway across the end wall of a room, thirty feet long from end to end. Halfway across the opposite wall there prowled a hungry spider in search of food.

His senses sharpened by appetite, the spider suddenly became aware of the sleeping fly at the very moment when he was one foot up from the floor. In eager anticipation of the feast, he scurried towards his prey by the shortest possible route.

The room, bare of windows and boasting only one narrow door, was twelve feet high and twelve feet wide. It was cold and cheerless, but the spider had a date with his fly. How far did he travel to reach his sleeping victim?

130. The Ballad of Ballygan

To his dismay
He learned next day
What havoc war had wrought:
He had at most
But half his host,
Plus ten times five two nought.

One-ninth were lain
In beds of pain,
With hundreds eight beside;
One-sixth were dead,
Enslaved or fled,
All lost in warfare's tide.

And so 'tis told
To warn the bold:
Aggression doesn't pay.
How many men
Marched with him, then,
At dawn the previous day?

131. The Art of Selling

Simple Simon's sister Sarah worked in a newsagent's stall for a few days after leaving school. A little book she found on the rack told her all she wanted to know about selling, and she faced the first customer with confidence.

'How much are those?' asked the man, pointing to an open box of gaily-banded cigars.

Sarah didn't hesitate. 'A cigar and a half per day and a half for a week and a half would cost a pound and a half,' she told him. That's what the book had said: 'Be different. Make them interested.'

And it worked, for he laughed and bought a pound's worth. How many cigars was that?

132. How High was the Window?

It takes a youngster to notice things. Out with his Dad, Ken suddenly pointed across the street. 'Look! There's something hanging down the front of that house.'

They crossed over to investigate and, sure enough, a fishing-line dangled down from a closed upper window. At the end of the line was a little lead sinker which just touched the concrete pavement.

'Now we'll find out how high that window-sill is,' said Ken's father. Taking a steel measuring-tape from his pocket, he pulled the sinker to one side until it was two feet above the pavement, keeping the line taut. In this position the sinker had been displaced horizontally exactly ten feet.

By that time Ken had lost interest in the whole affair, but he made a good show of pretending to understand the way his father worked out the height. How high do you make it?

133. First Snow of the Winter

It was early December. 'Cold, with snow,' had been the forecast, and this time it was right. From Theatre Royal right down to Allen Street there was a complete hold up of trams—some trouble with a 'learner' who had skidded into a lorry, shedding a wheel or two in the process.

'We'll get out,' Bill told his wife. 'It's far quicker walking.'

Sliding and slithering through the slush, Betty followed in his track as they struggled up High Street. For every pace forward, Bill slipped back a third of a pace; Betty, faring better with her shorter steps, slipped back only a quarter of every step forward. But Betty took five steps while Bill took three.

When Bill reached the Post Office he was surprised to find his wife still close at his heels. 'How d'you like it?' he shouted through the driving snow. 'Fine!' was her cheery reply. 'But you just keep on bull-dozing.'

Bill's forward strides averaged thirty inches. How long were Betty's?

134. That Expense Account

George and Harry were discussing their jobs. Although they sell for rival firms, they still remain the best of friends.

Today George was thoroughly disgruntled. 'I've had another row with them over expenses,' he complained, 'and it's always the same. They're downright mean.'

Harry smiled rather patronisingly. 'You haven't trained them right,' he told his friend. 'Now there's never any fuss with my people. Last week I spent as much on Monday as on Wednesday, on Tuesday two pounds more than the amount by which Thursday's expenses were less than Monday's, on Thursday five pounds less than on Friday, and on Friday, and on Wednesday I spent more than six pounds.

George was completely lost. 'My goodness!' he gasped. But Harry continued unperturbed. 'And the funny thing was that each day I spent an exact number of pounds, and the total for the week was twenty-nine pounds.'

Presumably Harry doesn't make out his account in such an obscure form, but it would be interesting to know what he spent on the Friday.

135. Sal and the Skirts

Sal and Sue are still employed as operators in the factory. Just recently they have been working on a special line of women's suits, Sue on the coats and Sal on the skirts. It was most monotonous work—eight hours a day, day after day, turning out garments with almost automatic regularity: always the same design, the same material and even the same size fitting. But the money was good, and the girls didn't complain.

Sal was making twelve skirts more in a day than the number of coats Sue made. The time it took Sue to make the number of coats by which the number of skirts Sal made in a day exceeded the number of coats Sue made in two days was the same as the time it took Sal to make two skirts fewer than the number of coats Sue made in a day.

So now we know how many skirts Sal made in a day, but do you?

136. The Five Threes

In this skeleton sum, two 3-figure numbers are multiplied together. There are only five threes, however, and these are shown; all the other figures are indicated by crosses. Now we have to find the two numbers and their product.

```
    x x 3
    x x 3
    -----
    x x 3
  x 3 x
3 x x
---------
x x x x x
```

137. A Trip to Neepawa

Ken was tired of sitting in the back of the car, watching his father's neck and wondering why it wasn't smooth like Mum's. 'What sort of a town is it?' he asked. 'Will I be able to see a movie tomorrow?'

'Neepawa's bigger than Roblin, where Uncle Pierre lives,' replied his father, 'but we'll have to see what Aunt Claire has fixed for tomorrow.'

But Ken was still curious: 'How much bigger?' he persisted. His father thought for a moment. 'This'll keep you busy for a bit,' he chuckled. 'If two hundred and fifty-four people moved from Neepawa to Roblin, Roblin would have half the population of Neepawa; but when we three arrive in Neepawa the population will become three times that of Roblin.' 'And just forget about other visitors,' he added, as an afterthought.

For a long time there was only the sound of muttering from the back of the car as Ken scribbled away, but the boy figured it out. What do you make the population of Neepawa at that time?

138. How Old is 'Old'?

Mr White loves kids, especially the noisy, rather grubby youngsters he meets going home in the street-car. The kids love him too, and always tell him the latest news from school.

Today they were full of their new teacher. When asked what she was like, they all agreed on one point: 'She's old!'

Amused by this, as he happens to know the girl who was an old school-friend of his youngest daughter, Mr White asked what they meant by 'old'. One said she must be twenty-seven, another suggested thirty-one, a third was sure she was thirty-six, and a fourth child went right up to thirty-nine.

In fact one guess was only one year out, one erred by three years, one was six years wrong, and one was as much as nine years from the truth. But could *you* give the teacher's age?

139. 'Silver Threads'

'Say, Dad, how old is Mum?' asked Ronnie. 'Jack says she's older than his mother, but Mum looks years younger to me.'

'Of course she does,' agreed his father, 'but don't tell Jack I said so.'

Mr Dale thought for a moment. 'You can work out your mother's age yourself. Add together the two figures of her age and multiply by the age of your younger sister, and you'll get what your Mum was a year ago.'

Jack's younger sister, Pam, certainly cannot be less than seven years old, but we'll leave you to work out the girl's age.

140. Doreen Got her Puppy

'Isn't he cute?' exclaimed Doreen, pointing to the spaniel pup in the window of the pet-shop, 'I want him, Daddy.'

Her father, indulgent as fathers are where small daughters are concerned, allowed himself to be led inside. They inspected the very mixed and somewhat dejected collection of cats and dogs and birds, but Doreen was still set on the spaniel.

'Well,' laughed her father, 'there are thirty-six animals here including the birds, and you can have your puppy if you tell me how many legs there are altogether.'

Doreen counted frantically and arrived at a total of exactly one hundred. She was far too young even to think of classing a human being as an animal, and so she got the pup. How many birds would you say there were in the shop?

141. So Many Oranges

'What a lot of oranges!' exclaimed their mother when the children laid out their purchases on the kitchen table. She turned to John, older and perhaps more responsible than his two sisters. 'What did they cost?'

'The girls bought them,' replied John. 'Jane got some and Anne got some also, but at the same price, and they spent sixty pence altogether.' He hesitated a moment and then added: 'If Jane had paid as many pence each as the number Anne bought, what she got would have cost ninety-six pence.'

'And each would have cost me an exact number of pence more than they did cost,' said Jane.

'You mean "it", not "they",' laughed her mother, but she was far too confused to count those oranges.

How many did the girls buy altogether?

142. The Marbles Rolled

If you want to reduce that waistline, just drop fifteen marbles on the floor of a swaying tram-car and then try to recover them without bending your knees. That's what Peter did, but he bent his knees—he has no waistline worry!

He had been admiring his treasures: red marbles, green marbles, blue marbles; seven of one shade, five of another, and three of the third colour. Then the tram gave a violent lurch and the whole lot went skittering over the floor. But Peter managed to recover all but four of them before he had to alight with his mother at their stop.

'How many of those pretty greens have you left?' asked his mother, genuinely sympathetic. 'I've got two more reds than greens,' he replied rather sulkily, 'but of one colour I've now only got two left.'

So how many green marbles did he have before the tragedy in the tram?

143. Late, as Usual

'You're late,' called out Betty from the frontdoor. 'Steve left some time ago. You know how he hates speeding, but he said you'd overtake him.'

Mike looked at his watch. 'Yes, I'll do that long before he gets to Polmood. If I averaged only thirty-two I'd catch him in three hours, and at thirty-six in only two.' Steve is one of those cautious drivers: he dodders along at what he calls his 'safe speed'. Mike knows the exact figure, and laughs at his friend—'As if any one speed was safe!'

However, Mike wasn't wasting any time today. He put his foot down and drove off in a cloud of dust. With a fairly clear road all the way he was able to maintain a steady forty miles an hour as he had expected. So how long did it take him to overtake Steve?

144. Figs for Four

Freddy was sent out to buy some green figs. 'You'll have to give me half the figs and also half a fig,' his mother told him, 'and then you'll have to give Agnes half the remaining figs and also half a fig, and after that Jackie will get half the balance and also half a fig.'

'But what do I get, Mum?' asked the boy after repeating the rigmarole to himself a few times.

'You know you don't like figs,' he was told, 'but you can arrange it so that there'll be just one fig over for yourself.'

And that's just what Freddy did. But how many figs did he buy?

145. A Lady and her Lingerie

His lady was lovely, with glamour galore.
She led him one day to a lingerie store.
A dozen of these and two dozen of those
Transparently feminine silk so-and-so's
Would cost him, they said, six bucks less, if you please,
Than a dozen of those and two dozen of these.
Such extravagance, though, was way out of his reach.
He'd only six bucks, which just bought one of each.
Ignoring the frowns and 'tut-tuts' of the grundies,
Just figure the price of each item of undies.

(*Note:* Sorry about this! A 'buck' is $1.)

146. Her Secret

'How old are you, Gran?' asked Ken politely. The old lady smiled at the little boy. 'That's a question no gentleman should ask,' she replied, 'but I'll tell you if you promise to keep it a secret.'

Ken nodded, and his grandmother went on: 'If you reverse my age you'll get half of what I shall be in a year's time.'

That kept him quiet for quite a while, but what do you make her age?

147. The Curse of Man

Bert burst out of the bathroom, blood on his chin, swearing he'd give up shaving. 'You'll do nothing of the sort,' laughed his wife, 'for you know what I think of beards. And besides it's your birthday, so stop cussing.'

Sitting down at breakfast some minutes later, Bert came back to the subject. 'I've shaved every day of my life since Dad gave me my first razor on my eighteenth birthday,' he remarked, 'and if I had shaved for six times as many months as I have years and for three times as many weeks as I have months I should be exactly five years older than I am today.'

This was not quite accurate, for he had forgotten about leap years, but maybe you can work out his age.

148. On the Road

'Another day off!' exclaimed Arthur, seeing the golf-bag in his friend's car. 'Some people have all the luck.'

'But I make up for it in the winter,' replied Tony who covers a wide area as a textile salesman, 'and I'll be working harder next month.' He paused a moment, and then went on: 'Last month I averaged five pounds for every day I was on the road, but for two of those days I made less than one pound altogether—can't remember the exact amount but it was an exact number of pence, no half pence.'

Arthur knows his friend's movements very well, and he certainly knew how many days Tony had worked the previous month. 'Apart from those two bad days, then, you must have averaged one penny over six pounds for the other few days you worked,' he commented enviously.

How much had Tony earned that month?

149. Three Friends and their Birthdays

Christmas 1968 was over and the three teen-agers were looking forward to their birthdays. Bella was born in February, while Betty and Beryl have their birthdays the previous month.

Beryl is just eight weeks older than Bella, but three days younger than Betty. What year were they born?

150. It's Collars Today

'Well, this is a fine time to come in to work!' cried Sue, greeting her friend who had just entered the busy work-room. 'Why not?' replied Sal. 'That's the best of piecework.'

Sue nodded. 'We're on collars today,' she remarked, pointing to a little heap beside her, 'and I've made thirteen already. But you've got a different style, and your rate will be twice as much as I get, you lucky girl.' 'That's fine,' said Sal, as they both settled down to their machines, 'I reckon you'll do seven of yours to my four.'

For the rest of that day the two girls worked hard and fast, stopping only to have a quick snack together at midday. When they finally stopped that evening, it turned out that Sal's estimate was correct; and, strangely enough, her earnings for the day were exactly the same as Sue's.

How many collars did Sal make?

TYPICAL SOLUTIONS

TYPICAL SOLUTION 'A'

26. Cigars for Dad

Say cheap cigars cost x pence each.

Then his first selection cost $(4x + 8x + 4x + 24)$ pence, so he had $(16x - 8)$ pence.

Medium-price cigars cost either $2x$ pence or $(x + 6)$ pence each.

Hence, $16x - 8 = 24x$ or $12x + 72$,

whence $x = -1$ or 20.

Cigars cannot cost negative price, so $x = 20$.

The prices were 20p, 26p, and 40p.

He spent £3·12.

TYPICAL SOLUTION 'B'

1. Murder on the River

He paddled at x miles per hour through the water.

The current was y miles per hour over the land.

After 15 minutes he was $\frac{x}{4}$ miles from the glove.

The glove took $\frac{1}{y}$ hours to reach the steps.

So, after he turned, the glove took $\left(\frac{1}{y} - \frac{1}{4}\right)$ hours,

i.e. $\frac{4 - y}{4y}$ hours, to reach the steps.

In that time he moved through the water $\frac{x(4 - y)}{4y}$ miles.

Continued

But we have seen he was $\frac{x}{4}$ miles from the glove.

So $\frac{x}{4} = \frac{x(4-y)}{4y}$, whence $y = 2$.

Hence the speed of the current was 2 miles per hour.

TYPICAL SOLUTION 'C'

138. How Old is Old?

As the maximum error was 9, and the difference between lowest and highest estimates was 12, the correct age was between 27 and 39.

Say the error of 27 was a years, the error of 39 was b years.

Then $a + b = 12$.

Errors were 1, 3, 6 and 9.

Hence we have $a = 3$, $b = 9$ OR $a = 9$, $b = 3$ whence the correct age must have been 30 or 36.

But 36 was one of the incorrect estimates, so the correct age was 30 years.

TYPICAL SOLUTION 'D'

4. The Spider and the Fly

Flies have 6 legs.

The spider would have required x females with 8 legs, y males with 3 legs.

Hence $8x + 3y = 28$.

Continued

This is the simplest form of 'indeterminate equation'—one equation with two unknowns.

Divide through by 3 (i.e. the smaller coefficient).

Then $$3x - \frac{x}{3} + y = 9 + \frac{1}{3}$$

Now x and y are whole numbers and therefore $\frac{x+1}{3}$ must be a whole number.

Say $\frac{x+1}{3} = t$ where t is any whole number, positive or negative or zero.

Then $$x = 3t - 1$$

Substitute this value for x in the original equation, so getting $$y = 12 - 8t$$

In this particular case we know that x and y must both be positive, and so we see that t must be 1.

This gives $$x = 2,\ y = 4.$$

Hence there would have to be 6 'flies like that': 4 males and 2 females.

TYPICAL SOLUTION 'E'

27. A Tale of the Tube

The ages are: Mum, x years; Bob, y years.

From data given, we have the equation:

$$5y^2 - 2xy + 3x + 6 = 0.$$

This is an 'indeterminate equation of the second degree': two unknowns, including the square of one or both. For the

Continued

first step we treat it as an ordinary quadratic, in this case in y:

$$y = \frac{x \pm \sqrt{(x^2 - 15x - 30)}}{5}$$

As x and y are whole numbers, the expression under the square-root sign must be a square, and positive.

Now let $$x^2 - 15x - 30 = k^2$$
where k is any whole-number that will satisfy the 'square' condition.

Treating the equation $x^2 - 15x - 30 - k^2 = 0$ as a regular quadratic, we get:

$$x = \frac{15 \pm \sqrt{(4k^2 + 345)}}{2}$$

Again, the expression under the square-root sign must be a square, and we have to find values of k that will make it so. There are various ways of dealing with this, one being by trying successive values. In this case there is an easier method.

$4k^2$ is a square, which simplifies matters.

Let $4k^2 + 345 = t^2$ where t is a whole number.

Then $$t^2 - 4k^2 = 345$$
i.e. $$(t + 2k)(t - 2k) = 345.$$

Taking the factors of 345, we have:

$$345 = 345 \times 1, \text{ or } 115 \times 3, \text{ or } 69 \times 5, \text{ or } 23 \times 15.$$

Now tabulate these alternatives, coupled with the expressions $(t + 2k)$ and $(t - 2k)$, where $(t + 2k)$ must be greater than $(t - 2k)$:

	$t + 2k =$	345	or 115	or 69	or 23
	$t - 2k =$	1	or 3	or 5	or 15
So	$4k =$	344	or 112	or 64	or 8
	$k =$	86	or 28	or 16	or 2
Making	$x =$	94	or 37	or 26	or 17
	$y =$	36	or 13	or 2	or 3

Continued

Now we can see that Mum could not be 94 if Bob were 36; and Bob cannot be 2 or 3.

Hence we are left with the one solution:

$$x=37,\ y=13.$$

So Bob was 13 years old.

TYPICAL SOLUTION 'F'

7. The Funny Figures one Sees

Let N be the required number.

$$N=3a+1=5b+2=7c+3=9d+4$$

where a, b, c and d are whole numbers.

Then $$7c-9d=1.$$

This is an indeterminate equation, but not of the simplest type. (See *Typical Solution 'D'*.)

Divide through by 9:

$$c-\frac{2c}{9}-d=\frac{1}{9},\quad \text{whence } \frac{2c+1}{9} \text{ is a whole number.}$$

Now multiply the numerator by 4. This multiplier is chosen as being the lowest multiplier that will make the new coefficient of c one less or one more than a multiple of 9.

We get $\frac{8c+4}{9}$, which must also be a whole number.

This becomes, $$c-\frac{c}{9}+\frac{4}{9},$$

whence $\frac{c-4}{9}$ is a whole number, say t.

Then $$c=9t+4.$$

Now $$5b+2=7c+3.$$

Substituting in this for c:

$$5b-63t=29.$$

Continued

By similar process this gives $b = 63k + 31$, where k is a whole number.

Again $$3a + 1 = 5b + 2.$$

Substituting in this for b,

$$3a - 315k = 156.$$

Simplifying this we get

$$a - 105k = 52$$

i.e. $$a = 105k + 52.$$

Now $N = 3a + 1$, and N is in the 'one hundreds'.

So k must be zero, whence $a = 52$.

That gives us: $N = 157$.

TYPICAL SOLUTION 'G'

8. Trust a Postman to Know

'Each dwelling faces its twin', so there must be an even number of houses, say $2n$ houses.

The distance between the first and last houses on each side will be $64(n - 1)$ feet, and the distance between each house and its twin across the road is 112 feet.

The diagrams show the alternative routes:

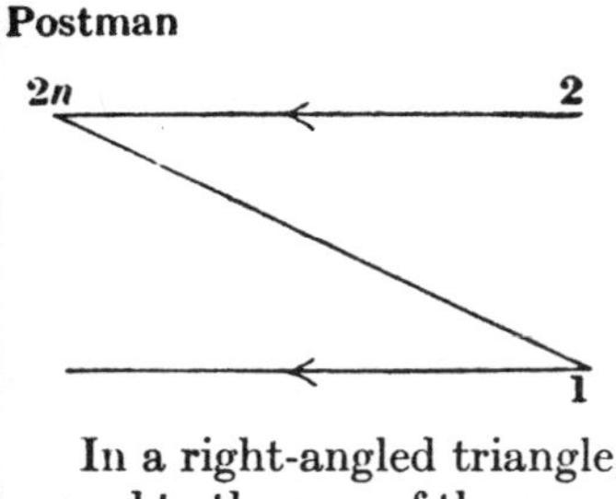

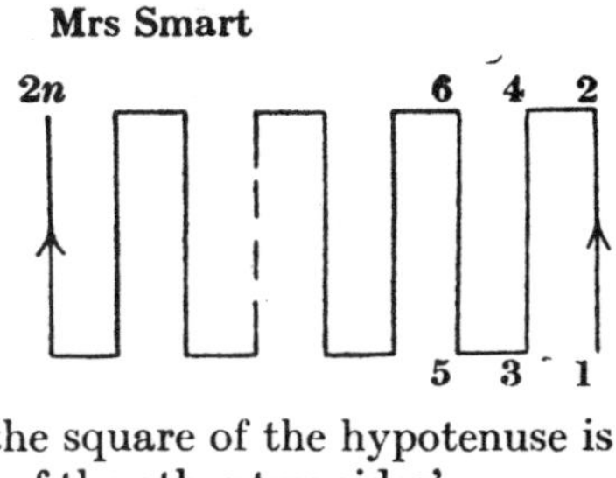

In a right-angled triangle, 'the square of the hypotenuse is equal to the sum of the squares of the other two sides'.

Continued

Hence the total distance by the postman's new route is:

$$(128(n-1)+\sqrt{\{64^2(n-1)^2+112^2\}})\text{ feet}$$

i.e.

$$16(8n-8+\sqrt{\{(16n^2-32n+65)\}})\text{ feet.}$$

And the total distance by Mrs Smart's route would be:

$$\{64(n-1)+112n\}\text{ feet}$$

i.e.

$$16(11n-4)\text{ feet.}$$

Equating these two distances, which are equal, and simplifying:

$$n^2-8n+7=0$$

whence

$$n=1 \quad \text{or} \quad n=7.$$

There are more than 2 houses, hence $n=7$.

So there are 14 houses in Myrtle Drive.

TYPICAL SOLUTION 'H'

10. No Seven Here

```
    2 x x
    3 x x
    -----
    5 x x
  x 4 x
x x 3
---------
x x x x x
```

The 1st product starts with 5, so the 1st multiplier must be

Continued

2. Inserting that figure, and since 3, the 3rd multiplier, gives a product ending in 3, we now have:

```
      2 A 1
      3 B 2
      -----
      5 x 2
    x 4 x
  x x 3
  ---------
  x x x x x
```

For ease of reference, letters *A* and *B* have been used.

As (2 *A* 1) multiplied by 2 gives (5 x 2), *A* is greater than 4.

Then, since the 2nd product has only 3 digits, *B* is less than 5.

Now consider the 2nd product. Within the limits which we have established for *A* and *B*, there are only two feasible alternatives:

$$B = 2,\ A = 7, \text{ or } B = 3,\ A = 8.$$

But there can be no 'seven'; so $B = 3$, $A = 8$.

Hence we are left with:

$$281 \times 332 = 93292.$$

TYPICAL SOLUTION 'I'

11. Tim and his Brother Tom

Tom's age is x years, Tim's y years.

Working backwards from the statement of fact (i.e. 'Tom *was* two years younger etc.'), we get:

When Tim is twice his present age, Tim will be $2y$.

Continued

When Tom was 15 years younger than that, Tom was $(2y-15)$.

When Tim was twice as old as that, Tim was $(4y-30)$.

When Tom was as old as that, Tom was $(4y-30)$.

Working backwards from the end, we get:

7 years ago, Tom was $(x-7)$.

When Tim was 1 year older than that, Tim was $(x-6)$.

When Tom was a third of that, Tom was $\frac{x-6}{3}$.

When Tim was 3 years older than that, Tim was $\frac{x+3}{3}$.

When Tom was 2 years younger than that, Tom was $\frac{x-3}{3}$.

So we have $$4y-30=\frac{x-3}{3}$$

and also $$x+y=17.$$

Whence $$x=9,\ y=8.$$

So Tim is 8 years old.

TYPICAL SOLUTION 'J'

13. No Catch in This

Let the three digits be x, y, and z in that order.

Then $$3z-2x+12y=66.$$

This is an indeterminate equation with three unknowns; in most cases this is handled on much the same lines as an

Continued

indeterminate equation with two unknowns. In this case, however, there is a shorter method of solving it.

$$12y = 66 - (3z - 2x)$$

and $3z - 2x$ cannot be negative, hence y is less than 6.

None of the digits can be 5, so y is less than 5.

$3z - 2x$ cannot exceed 25, i.e. $27 - 2$.

So $12y$ cannot be less than 41, whence y is greater than 3.

So $$y = 4.$$

Thus $$3z - 2x = 18.$$

This indeterminate equation can be solved at sight:

$$x = 0,\ z = 6 \text{ or } x = 3,\ z = 8.$$

The wording of the first verse, coupled with the assurance that 'there's no catch in this', suggests that the first digit cannot be zero.

Hence the digits are 348.

TYPICAL SOLUTION 'K'

14. Kim's Camera

The camera cost him exactly £y.

He paid x pounds in notes, and x coins of each of 50p, 10p, 5p, 2p, 1p denominations.

Altogether he paid $x(100 + 50 + 10 + 5 + 2 + 1)$ pence.

Hence $168x = 100y$, so $42x = 25y$, and y is less than 50.

But x and y are whole-numbers, hence $x = 25$, $y = 42$.

The camera cost him £42.

TYPICAL SOLUTION 'L'

15. The Old Soldier

He enlisted in 1890, so the date in question was not before that year.

Say the digits of the date were 1*ABC* in that order. This has actual value $1000 + 100A + 10B + C$.

Exchange of 2nd and 4th digits would give a new date, 1*CBA*, with actual value $1000 + 100C + 10B + A$.

Hence, as the new date is the later of the two,

$(1000 + 100C + 10B + A) - (1000 + 100A + 10B + C) = 99.$

Whence $C - A = 1.$

Now A must be 8 or 9.

If $A = 9$, C would equal 10, which is impossible.

So $A = 8$ and $C = 9$.

The date was not before 1890, so $B = 9$.

Hence the date was 1899.

TYPICAL SOLUTION 'M'

20. How Many Marbles?

They started with: Tam, $10x$; Tim, x; Tom, $2x$.

When Father came out, Tim had won a marbles, and either Tam or Tom had lost a.

The boy who grumbled could not have been Tim or Tom, so he must have been Tam. This is confirmed by the term 'third boy' as applying to Tim.

So it was Tam who had lost a marbles to Tim.

Continued

When Father came out, the position was:

Tam: $10x - a$.
Tim: $x + a$.
Tom: $2x$.

So when the game ended, they had:

Tam: $2x$.
Tim: $10x - 3$.
Tom: $(13x - \{2x + (10x - 3)\}) = (x + 3)$.

But two of the boys 'ended up equal'.

So $2x = (10x - 3)$, which is impossible
or $10x - 3 = x + 3$, which is impossible
or $2x = x + 3$, which is quite possible.

Hence, there being no other alternatives, $x = 3$.

So they had 39 marbles among them.

TYPICAL SOLUTION 'N'

24. Those Licence Plates

Represent their respective numbers, in conformity with the information, as:

Paul: $A\ B\ C\ D$. Dick: $A\ D\ D\ E$. Hal: $D\ B\ C\ A$.

Giving these their true values, from Hal's statement we get:

$$1000A + 100B + 10C + D + 1000A + 110D + E$$
$$= 1000D + 100B + 10C + A.$$

Whence $1999A - 889D + E = 0$.

This is an 'indeterminate equation' with three unknowns, to be dealt with on somewhat similar lines to one with two unknowns.

Continued

Dividing by 889, we have:

$$2A+\frac{221A}{889}-D+\frac{E}{889}=0.$$

Whence $E=889t-221A$ $D=2A+t$.

Now if $A=0$ and $t=0$, then Dick's number would be 0000, and Paul and Hal would have the same numbers: this would have been an amazing coincidence and could not have escaped mention. So we may exclude that possibility.

So t must have minimum value 1 and must be positive.

From $D=2A+t$, since D must be less than 10, A cannot be greater than 4.

From $889t=221A+E$, since A must be 1, 2, 3, or 4, and E can be any digit from 0 to 9, t cannot be greater than 1.

Hence $t=1$.

Thence we have:

$$E=889-221A,\ D=2A+1.$$

E and D are both less than 10, so $A=4$.

Whence $E=5$ and $D=9$.

So Dick's number was 4995.

TYPICAL SOLUTION 'O'

28. The Beauties of Kalota

Kassa said: Kissa, 23; Kessa, 22.
Kessa said: Kossa, 20; Kissa, 22.
Kissa said: Kassa, 22; Kossa, 19.
Taking the three together:
If Kissa was 23, then she was not 22.

Continued

So in that case Kossa was 20, and, as Kossa could not then be 19, Kassa was 22.

This would make Kessa's age 19.

If Kessa was 22, then Kissa was not 22 (i.e. there are no twins).

So in that case Kossa was 20, which again makes Kassa's age 22.

This is impossible, as there are no twins.

So Kissa was 23, and the four ages were:

Kissa: 23
Kossa: 20
Kassa: 22
Kessa: 19.

TYPICAL SOLUTION 'O': PART TWO

29. The Beauties of Kalota

By Boolean Algebra

Many problems of this type may be solved very neatly by the methods of Boolean Algebra, the basis of the mathematics of logic. Only a bare outline of the elements is possible here, but it may be of interest.

We adopt the convention that something 'true' has the value 1, and something 'false' the value 0. Using code symbols for the 'somethings', we can then form expressions and equations which may be treated much the same as those in normal algebra. A very simple example will show how the method is used.

Say we have two statements about the name of a girl: one man said 'Betty Price', the other 'Gwen Price', and we are told that each made one mistake. Obviously Price was her

Continued

surname, but her first name could not have been either Betty or Gwen. Now let us see how this would be handled by Boolean Algebra.

We have only the two numerical 'values', 0 and 1. There's nothing more true than 'true': if, in the course of the working, we arrive at any number greater than unity, we must represent it as unity.

Let B stand for Betty, G for Gwen, and P for Price. Then we can represent each statement in two ways:

Multiplication. If both B and P were equal to 1 (i.e. true), the product $BP=1$. But if either B or P has value 0 (i.e. false), the product $BP=0$.

Addition. If either B or P (or both) has value 1, (i.e. is true), then $B+P=1$.

So we can say, $(B+P)(G+P)=1\times 1=1$

whence $BG+BP+GP+P^2=1$

But $BP=0$, $GP=0$, and obviously $BG=0$, so we are left with $P^2=1$, i.e. $P=1$, which tells us that her surname was Price. As $P=1$, and $BP=0$ and $GP=0$; we have $B=0$ and $G=0$, which tells us that her first name was neither Betty nor Gwen.

Now we can deal with the teaser of the Kalotan girls.

Let our code be:

KISSA—I	19—a
KESSA—E	20—b
KOSSA—R	22—c
KASSA—A	23—d.

In code, then: Kassa said: Id, Ec
Kessa said: Rb, Ic
Kissa said: Ac, Ra

So we have:

$$Id+Ec=1 \quad \text{and } Id.Ec=0$$
$$Rb+Ic=1 \quad \text{and } Rb.Ic=0$$
$$Ac+Ra=1 \quad \text{and } Ac.Ra=0$$

Continued

There are no twins, so any term of the form $Ec.Ic$ must have 'value' zero.

No girl can have two ages, so terms of the form $Id.Ic$ must have 'value' zero.

Now combine as $(Id + Ec)(Rb + Ic) = 1$

Whence $Id.Rb + Id.Ic + Ec.Rb + Ec.Ic = 1$

Striking out terms with zero 'value', this becomes

$$Id.Rb + Ec.Rb = 1$$

Now bring in the third equation, as

$$(Id.Rb + Ec.Rb)(Ac + Ra) = 1$$

Whence

$$Ac.Id.Rb + Ac.Ec.Rb + Ra.Id.Rb + Ra.Ec.Rb = 1$$

Striking out terms with zero 'value', we are left with

$$Ac.Id.Rb = 1.$$

This shows that the ages were:

Kassa: 22
Kissa: 23
Kossa: 20.

So Kessa was 19.

ANSWERS

ANSWERS

(The Capital Letter in brackets indicates the Typical Solution which may provide a clue to the method of solving the problem)

1. (B) 2 m.p.h. 2. (A) 23 cars had passed.
3. (C) Uncle Jim had 38p in his hand.
4. (D) 4 males, 2 females. 5. (E) 5 oranges.
6. (A) 2 years older. 7. (F) No. 157.
8. (G) 14 houses. 9. (A) 10 cigarettes.
10. (H) 93292. 11. (I) Tim is 8 years old.
12. (A) 36 chickens. 13. (J) 348.
14. (K) £42. 15. (L) 1899. 16. (A) £1·94.
17. (D) 12 bicycles, 3 horses, 6 cars.
18. (A) She had spent £7·90.
19. (I) Peg is his niece. 20. (M) 39 marbles.
21. (O) Alfred and Betty Cross, Charles and Amy Briggs, Herbert and Clare Armstrong.
22. 1, 3, 9, and 27 kalens. The theoretical solution involves advanced application of the Theory of Numbers.
23. (A) 480 cars going north. 24. (N) No. 4995.
25. (D) Susie earned £18·70.
26. (A) He spent £3·12. 27. (E) 13 years old.
28. (A) Jim killed 4 birds.
29. (O) Kissa, 23; Kassa, 22; Kossa, 20; Kessa, 19.
30. (A) Milly made 14 muffins.
31. (B) 240 kilometres per hour.
32. (A) The area was 84 square yards.
33. (D) St Mark's, 27 seconds; St Mary's, 17 seconds.
34. (A) Mike paid £32·89. 35. (A) Price was 1s. 8d.
36. (I) Jim is 35 years old. 37. (A) The year was 1952.
38. (D) At 6.36.
39. (D) 5 @ 89p, 12 @ £1·24.
40. (O) Tuesday, June 10th. 41. (M) Anne was Bob's wife.
42. (B) Roy had travelled 24 metres.
43. (A) He had 69p altogether.
44. SNOW was 1728 (i.e., 828 + 828 + 72).

45. (H) 45353.

46. (A) Mike's normal pace is 30 inches.

47. The frontage was x yards. Then the depth was $\frac{120 - 4x}{3}$ yards. The depth being an exact number of yards, x must be a multiple of 3, say $x = 3k$. Then the area was $12k(10 - k)$ square yards, and k must be less than 10. Tabulating for whole-number values of k, from $k = 1$ to $k = 9$, we find that the maximum area is 300 square yards when $k = 5$. So the area of the plot was 300 square yards. (*Note:* This can be solved more readily using elementary calculus.)

48. (D) He had spent £2·28.

49. (A) 11 birds.

50. (A) The entry of 50p was wrong.

51. (D) 48 geese.

52. (A) The kettle would have cost £4·25 alone.

53. (A) He bought 6 ounces altogether.

54. (A) 102 miles.

55. (A) 6 children.

56. (O) Elsie was not at the party.

57. (J) No. 9317.

58. (A) 72 stamps.

59. 5 becomes 80, 15 becomes 74, 20 becomes 71. The rule of transformation obviously operates in proportion: (20 + 10) becomes (71 − 6), so 30 becomes 65.

60. (L) The correct change was £1·81.

61. If Jim bought 9, the outlay would be £3·80.
If Joe bought 9, the outlay would be £3·60.
So Jack bought 9, making the outlay £4.

62. (B) 12 miles.

63. (A) Doreen was born in 1939.

64. (G) The answer was 'Black and Red'.

65. (J) The number was 892.

66. (O) Mr Penn, lawyer; Mr Law, dentist; Mr Bankes, author; Mr Tooth, banker.

67. (C) The weight was 23 ounces.

68. 29786 + 850 + 850 = 31486.

69. Any one of the boys bought x $\frac{1}{2}$-litre and y litre cartons,

so $x + 2y = 5$. There are three solutions to this:

$$x = 1 \text{ or } 3 \text{ or } 5$$
$$\text{with } y = 2 \text{ or } 1 \text{ or } 0.$$

So each solution gives the purchase of one of the boys. Tom bought the most ½-litres, so he bought 5.

70. (E) Tony was 10 years old.

71. (A) Tom started with 19 cigarettes.

72. (O) John Everard, Jim Lennox, Jack Best.

73. (A) The tie cost £1·25.

74. (D) They had concealed 73 eggs.

75. (A) Gwen started with £6·50.

76. (B) Don's speed was 27 miles per hour.

77. (D) There were 38 ten-pences (11 children shared the money.

78. (I) He died aged 12 years.

79. (A) It was 28 feet high at the end of the 10th year.

80. (G) He had to travel 2½ metres.

81. (D) 7 cats ate 'mousie meat'.

82. (A) 209 voted for Kar.

83. (L) Simon paid £1·65 altogether.

84. (H) $813 + 518 = 1331$.

85. (A) He had £1·75 when he entered the first shop.

86. (N) His number was 2857.

87. (O) King Kormon died in the year of Karan.

88. (I) The rope was 20 feet long (the monkey weighed 5 kilos.).

89. (A) The weights were 4, 5 and 9 kilos.

90. (D) She bought 25 quarto and 13 foolscap size files.

91. Amelia is x, but feels y years old. From her final remark, Sally looks $(x - 20)$ years old. If Amelia had been $(x - 20)$ ten years ago, she would now be $(x - 10)$ years old. If Sally were as old as Amelia feels, Sally would be y years old, which is $\{(x - 10) + 20\}$ years old. So $y = x + 10$. Hence Amelia feels 10 years older.

92. (A) Aunt Emily was 42 years old.

93. (I) Susan is 20 years old.
94. (A) Bill dug 56 metres.
95. (A) She bought 18 @ 19p each.
96. (A) 60 pupils altogether. 97. (I) Jack is 11 years old.
98. (A) Tony's birthday was 16th July.
99. Ted lost £2 change, plus what the shoes had cost him.

100. (E) Kurtis's age, when his son was born, was x years; Kalia had been with him y months. Then we get

$$5x - \frac{y(y+1)}{200} = 23.\text{ Thence, } x = 55.$$

101. Mike started with x pence and won n games. If they played 7 games, $x = \frac{128 \times 555}{(128 - 3^n)}$ and, as $(128 - 3^n)$ is positive, n is less than 5. By trial, $(128 - 3^n)$ is not a factor of (128×555), so they did not play 7 games. So they played 6 games, and $x = \frac{64 \times 555}{(64 - 3^n)}$ with n less than 4. By trial $(64 - 3^3)$ is a factor of $(64 + 555)$, so $x = 960$. Hence Mike started with £9·60, ending with £4·05.

102. (A) Gwen's age was 19 years.
103. (O) Larry was first, Steve second, and Ken third.
104. (A) 8 brothers and 4 sisters.
105. (B) The distance was $4\frac{1}{2}$ kilometres.
106. (A) 48p in the purse.
107. (D) She bought 6 saucers.
108. (I) John is 8 years old.

109. There may be several alternative expressions for any number in this type of teaser. Examples are:

$$7 = 1 \times (4 + \sqrt{9}) = 1 \times (4 + 3)$$
$$15 = (1 + 9) \div \sqrt{0{\cdot}\dot{4}} = 10 \div \tfrac{2}{3}$$
$$28 = 4! + 1 + \sqrt{9} = (4 \times 3 \times 2 \times 1) + 1 + 3$$
$$30 = \Sigma(4) \times 1 \times \sqrt{9} = (4 + 3 + 2 + 1) \times 1 \times 3$$

110. (H) 'TWEED' must be 10772.

111. (D) The taxi fare was 54p.

112. The question could be: 'If, in my next question, I were to ask you, "Does the left road lead to Knokado?" would you answer "Yes"?' Then, if in fact the left road does lead to Knokado, she must say 'No'; and if it does not lead to Knokado, she must say 'Yes'. Remember, she could not lie in reply to two consecutive questions.

113. (O) Fred Leary, with 3 sons.

114. (L) & (N) The number has digits x, y, a and b in that order. So $1999x + 190y - 80a - 998b = 39$. Dividing by 10, we get: $x = 10t + 2b + 1$, t being a whole number.

Now x and b are both less than 10, so if

$$b < 5,\ x = 2b + 1$$

and if

$$b > 4,\ x = 2b - 9$$

Say $b < 5$, and substitute for x in the equation. Then we get

$$300b + 19y - 8a = -196.$$

Here b must be 0, 1, 2, 3, or 4 with $8a - 19y = 196$, 496, 796, 1096 or 1396.

But $8a - 19y$ cannot exceed 72, so none of these is acceptable.

Say $b > 4$. Then $300b + 19y - 8a = 1803$.

Then b must be 5, 6, 7, 8, or 9 with $19y - 8a = 303$, 3, -297, -597, or -897.

But $19y - 8a$ cannot exceed 171, so $b = 6$, $19y - 8a = 3$.

Hence, $y = 1$, $a = 2$. Also, $x = 2b - 9$, so $x = 3$.

Hence the number was 3126.

115. The ages of brothers are x and $(x + 10)$ years, and Sue's age is y years. Then $y + 5x = 42$, or $y + 5x = 52$, x and y being whole numbers as it was their birthday.

Also,

$$y = x + 8, \text{ or } y = x + 18,$$

so

$$x = y - 8, \text{ or } x = y - 18.$$

Substituting these values for x in the two original alternative equations, we get $6y = 82$, 92, 132 or 142.

A whole-number value for y is obtained only from $6y = 132$, whence $y = 22$. So Sue was 22 years old.

116. (D) Simon spent £1·12.
117. The speaker would be 63 years old.
118. (E) Pam was 10 years old.
119. (A) She had spent £6·20.
120. (E) 3 had already subscribed a total of £1·50.
121. (A) She still had £9.
122. (D) Aunt Amelia was 43 years old.
123. (B) He had walked for 26 minutes.
124. (O) Tam, Monday; Tim, Tuesday; Tom, Friday.
125. (A) The wall was 10 feet long.
126. (E) The number was 2205.

127. Weigh any 3 against any 3. *If they balance*, the 'fake' is the lighter of the other 3. Weigh one of these against one: if they balance, the 'fake' is the odd coin; if they do not balance, it is the lighter of the two being weighed. *If they do not balance*, the 'fake' is the lighter coin amongst the lighter 3. Proceed with these lighter three as in the previous case.

128. (I) He was 14 years old.
129. (G) The distance was 40 feet.

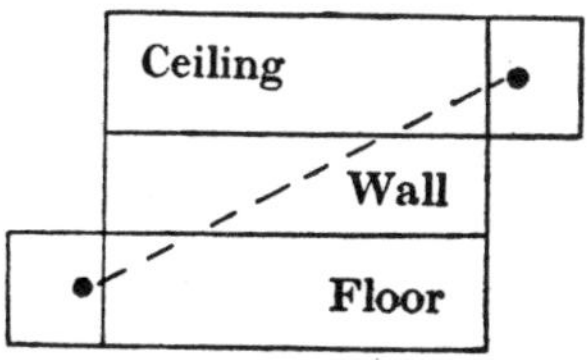

130. (A) 27,000 marched the previous day.
131. (A) 7 cigars for £1. 132. (G) 26 feet.
133. (A) Betty's stride was 16 inches.
134. (D) He spent £6 on Friday.
135. (A) 21 skirts in a day. 136. (H) $123 \times 163 = 20049$.
137. (A) Population of Neepawa was 2292.
138. (C) She was 30 years old.

139. (E) His mother was 41, and Pam was 8 years old.

140. (A) 22 birds. 141. (E) 20 oranges altogether.

142. He ended up with x red, $(x - 2)$ green, $(13 - 2x)$ blue. $(13 - 2x)$ cannot be 2; and x is not 2, as he did not have more than 7 of one colour. So $x - 2 = 2$. Hence he ended up with 4 red, 2 green, and 5 blue. So he started with 5 red and 7 blue, or 7 red and 5 blue. Either way he started with 3 green.

143. (B) Mike overtook Steve in $1\frac{1}{2}$ hours.

144. (A) He bought 15 figs. 145. (A) $2.75 and $3.25.

146. (L) She was 73 years old.

147. (A) Bert was 44 years old that day.

148. (D) Tony had earned £55.

149. They must have been born in a leap year: Betty on 1st January; Bella, 29th February. The only leap year between 1949 and 1955, was 1952: the year they were born.

150. (A) Sal made 52 collars.